**Molten Carbonate
Fuel Cells**

*Edited by
Kai Sundmacher, Achim
Kienle, Hans Josef Pesch,
Joachim F. Berndt, and
Gerhard Huppmann*

## 1807–2007 Knowledge for Generations

Each generation has its unique needs and aspirations. When Charles Wiley first opened his small printing shop in lower Manhattan in 1807, it was a generation of boundless potential searching for an identity. And we were there, helping to define a new American literary tradition. Over half a century later, in the midst of the Second Industrial Revolution, it was a generation focused on building the future. Once again, we were there, supplying the critical scientific, technical, and engineering knowledge that helped frame the world. Throughout the 20th Century, and into the new millennium, nations began to reach out beyond their own borders and a new international community was born. Wiley was there, expanding its operations around the world to enable a global exchange of ideas, opinions, and know-how.

For 200 years, Wiley has been an integral part of each generation's journey, enabling the flow of information and understanding necessary to meet their needs and fulfill their aspirations. Today, bold new technologies are changing the way we live and learn. Wiley will be there, providing you the must-have knowledge you need to imagine new worlds, new possibilities, and new opportunities.

Generations come and go, but you can always count on Wiley to provide you the knowledge you need, when and where you need it!

William J. Pesce
President and Chief Executive Officer

Peter Booth Wiley
Chairman of the Board

# Molten Carbonate Fuel Cells

Modeling, Analysis, Simulation, and Control

*Edited by*
*Kai Sundmacher, Achim Kienle, Hans Josef Pesch,*
*Joachim F. Berndt, and Gerhard Huppmann*

WILEY-VCH Verlag GmbH & Co. KGaA

**The Editors**

*Prof. Dr. Kai Sundmacher*
Max-Planck-Institut für Dynamik
komplexer technischer Systeme
Sandtorstr. 1
39106 Magdeburg
Germany

*Prof. Dr. Achim Kienle*
Max-Planck-Institut für Dynamik
komplexer technischer Systeme
Sandtorstr. 1
39106 Magdeburg
Germany

*Prof. Dr. Hans Josef Pesch*
Universität Bayreuth
Lehrstuhl für Ingenieurmathematik
Universitätsstr. 30
95440 Bayreuth
Germany

*Dipl.-Ing. Joachim F. Berndt*
IPF Beteiligungsgesellschaft
Berndt KG
Carl-Benz-Str. 6
68799 Reilingen
Germany

*Dipl.-Phys. Gerhard Huppmann*
MTU CFC Solutions GmbH
81663 München
Germany

*Cover Illustration*
Cover photograph with kind permission
from MTU CFC Solutions GmbH, Munich,
Germany.

**Library of Congress Card No.:** applied for

**British Library Cataloguing-in-Publication Data**
A catalogue record for this book is available
from the British Library.

**Bibliographic information published by
the Deutsche Nationalbibliothek**
Die Deutsche Nationalbibliothek lists this
publication in the Deutsche National-
bibliografie; detailed bibliographic data are
available in the Internet at ⟨http://dnb.d-nb.de⟩.

© 2007 WILEY-VCH Verlag GmbH & Co.
KGaA, Weinheim

**Typesetting**  Asco Typesetters, Hong Kong
**Printing**  Strauss GmbH, Mörlenbach
**Binding**  Litges & Dopf Buchbinderei
GmbH, Heppenheim
**Wiley Bicentennial Logo**  Richard J. Pacifico

Printed in the Federal Republic of Germany
Printed on acid-free paper

**ISBN:** 978-3-527-31474-4

# Contents

*Molten Carbonate Fuel Cells.* Edited by Kai Sundmacher, Achim Kienle, Hans Josef Pesch,
Joachim F. Berndt, and Gerhard Huppmann
Copyright © 2007 WILEY-VCH Verlag GmbH & Co. KGaA, Weinheim
ISBN: 978-3-527-31474-4

# Preface

Fuel cells generate electrical energy by electrochemical oxidation of chemical substances such as hydrogen, carbon monoxide, methanol, ethanol, glucose or other hydrocarbons. Due to their functional principle, fuel cells can achieve much higher efficiencies for energy conversion than conventional systems which are based on the Carnot cycle. Because of their high conversion efficiency, fuel cells will play a major role in the future mix of power supply systems.

The importance of fuel cells is also reflected by an exponential increase of journal papers and book contributions being published during the last two decades (see Fig. 1). Among the published papers and books, most are focused on polymer electrolyte fuel cells (PEMFC), direct methanol fuel cells (DMFC) and solid oxide fuel cells (SOFC). In comparison to these types of cells, the molten carbonate fuel cell (MCFC) so far has attracted relatively little attention. But this is in total contrast to the current status of system development. While large-scale applications of PEMFC, DMFC and SOFC-systems up to now are still quite rare, more than 20 demonstration plants of the MCFC HotModule type (power range: 250–300 kW) were already installed successfully for various applications by the company CFC Solutions Ltd./Ottobrunn in Germany.

As another trend, the literature analysis clearly reveals that the proportion of publications dealing with the model-based analysis and control of fuel cells is steadily increasing. But this research field is still young and therefore it comprises only about one tenth of the overall number of articles and books in this whole area (see Fig. 1). Designing efficient fuel cell stacks not only requires suitable electrode and membrane materials, but also powerful engineering methodologies and tools. Due to the complex nonlinear behaviour of fuel cells, their design and operation cannot be based on pure intuition. This is why advanced model-based methods for the analysis, control and operation have to be further developed in the next few years.

A comprehensive volume covering all aspects of model-based analysis, control and operation of fuel cell systems is still missing. To fill this gap, the present book was prepared with special focus on the MCFC as an example of high technical relevance. The presented concepts and methods are also transferable to other fuel cell types such as the PEMFC.

*Molten Carbonate Fuel Cells.* Edited by Kai Sundmacher, Achim Kienle, Hans Josef Pesch,
Joachim F. Berndt, and Gerhard Huppmann
Copyright © 2007 WILEY-VCH Verlag GmbH & Co. KGaA, Weinheim
ISBN: 978-3-527-31474-4

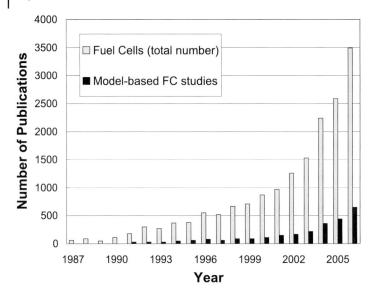

**Fig. 1** Journal publications on fuel cells from 1987 to 2006 according to the Science Citation Index.

The book is divided into three parts: Part A surveys the design and operation of MCFC fuel cells with special focus on the HotModule type, which integrates the generation of hydrogen by methane steam reforming and the electrochemical oxidation of hydrogen in one single unit. Part B is dedicated to process analysis by means of mathematical models describing the complex interactions of mass, energy and charge transport phenomena within the fuel cell stack. Part C focuses on how process models can be used for state estimation, for advanced control strategies and for solving steady-state as well as dynamic optimization tasks.

**Part A: Design and Operation**

Chapter 1 being written by Gerhard Huppmann, who is one of the inventors of *MTU's Carbonate Fuel Cell HotModule*, is concerned with the basic concepts and the key features of the cell design, presents possible applications using natural gas and other fuels, and discusses important economical aspects. Focus in Chapter 2 is on *Operational Experiences* which are reported by Koch et al. The authors collected a series of experimental data at different load scenarios at a 250 $kW_{el}$ MCFC HotModule stack which is installed as part of IPF's power plant at the University hospital in Magdeburg. The data form the basis for the identification of realistic model parameter values being a major prerequisite for reliable fuel cell simulations which are reported in the subsequent chapters.

**Part B: Model-Based Process Analysis**

In Chapter 3, Heidebrecht et al. present a rigorous *MCFC reference model* which accounts for the most important physico-chemical phenomena within the cell and also for the special recycle structure of the HotModule process. The model consists of a coupled set of hyperbolic and parabolic partial differential equations along with further ordinary differential and algebraic equations, completed by an integral equation expressing the overall charge conservation condition. The numerical treatment of these model equations, using finite volume discretization methods, results in a set of differential-algebraic equations (DAE) whose *index analysis* is performed by Chudej et al. in Chapter 4.

As outlined by Gundermann et al. in Chapter 5, *parameter identification* at an industrial-scale MCFC stack requires a special stepwise strategy which has to account for measurement errors as well as for possible leakages at the plant. With regard to cell dynamics, the solid heat capacity is the key parameter to be determined. Once realistic fuel cell model parameters have been identified, one can simulate the steady-state behaviour, particularly the current-voltage performance curve. Furthermore, the *process* analysis presented by Heidebrecht et al. in Chapter 6 includes the dynamic cell voltage decline after a load increase which can occur when the fuel cell is moved from a low-current operating point to a higher current load.

In Chapter 7, Krasnyk et al. present theoretical predictions of *hot spot formation and steady-state multiplicities* in high-temperature fuel cells. The discovered phenomena are relevant for the MCFC as well as for the SOFC and arise from the increase of the ion conductivity of electrolytes at increasing operating temperatures.

In the final chapter of part B, Heidebrecht et al. analyse and compare different *conceptual designs and reforming concepts* by means of a simple along-the-channel anode model. From this study it turns out clearly that the highest current yields are attainable by direct internal reforming within the anode channels.

**Part C: Optimization and Advanced Control**

The efficient operation of industrial scale fuel cell plants such as the MCFC Hot-Module requires continuous monitoring of key state parameters. In particular, for safe load changes one would like to know the spatially distributed temperature field within the fuel cell stack. But the experimental accessibility of internal temperatures is quite limited. Therefore, as a very promising monitoring alternative, Grötsch et al. in Chapter 9 show how to develop a model-based observer via *model reduction and state estimation* for tracing the dynamic evolution of the MCFC temperature field.

The subject of Chapter 10 which was written by Sternberg et al. are *optimal control strategies* for load changes between two predefined steady-state operating points. The aim of control is to attain the new steady state as fast as possible after load change. This leads to a so-called boundary control problem which can be tackled with the help of the software package NUDOCCCS.

In the final contribution, Heidebrecht et al. show that the *optimization of the reforming catalyst distribution* can lead to significant improvements of the fuel cell efficiency. Since methane steam reforming is an endothermic reaction, it can be used as an internal cooling element for the exothermic electrochemical reactions.

In the appendix, the complete set of equations and related parameters is given for the *MCFC reference model* which is the "mother" of all reduced model variants being used for simulation, analysis, control and optimization as discusses in preceding chapters. This information will serve as a source of information for readers who are interested to get all details in order to start their own comparative studies.

## Book History and Acknowledgements

The present book presents the outcome of the joint research project "Optimised control of fuel cell systems using methods of nonlinear dynamics" which was performed from 2002 until the end of 2005 in close collaboration of five German research groups from academia and industry. The project was coordinated and organized by the editors of this book and their colleagues at the Otto-von-Guericke-University Magdeburg, the Max-Planck-Institute for Dynamics of Complex Technical Systems in Magdeburg, the University of Bayreuth, the power plant operating company IPF Heizkraftwerksbetriebsgesellschaft Ltd., and the Molten Carbonate Fuel Cell producing company MTU CFC Solutions Ltd. in Ottobrunn. The financial support of the joint project from the Federal Ministry of Education and Research (BMBF) in Germany is very gratefully acknowledged. Intermediate results were discussed in 2004 on a MCFC workshop with financial support from the German Competence Network Pro3 e.V. which is also gratefully acknowledged.

The editors like to thank particularly their colleagues Peter Heidebrecht, Matthias Gundermann, Michael Mangold, Markus Grötsch, Mihai Krasnyk, Kurt Chudej, Kati Sternberg and Mario Koch for their excellent support during the project work and in preparing the manuscripts being the basis for the present book publication. Last but not least, we are very thankful to Dr. Rainer Münz from Wiley-VCH for his helpful assistance during the production of this book.

March 2007
*Kai Sundmacher, Achim Kienle, Hans-Josef Pesch,*
*Joachim Berndt, Gerhard Huppmann*

# List of Contributors

**Joachim F. Berndt**
IPF Beteiligungsgesellschaft
Berndt KG
Postfach 1110
68795 Reilingen
Germany
joachim.berndt@ipf-online.com

**Kurt Chudej**
Lehrstuhl für
Ingenieurmathematik
Universität Bayreuth
Universitätsstraße 30
95440 Bayreuth
Germany
kurt.chudej@uni-bayreuth.de

**Markus Grötsch**
Max-Planck-Institut für Dynamik
komplexer technischer Systeme
Sandtorstraße 1
39106 Magdeburg
Germany
groetsch@mpi-magdeburg.mpg.de

**Matthias Gundermann**
Process Systems Engineering
Lehrstuhl für Systemverfahrenstechnik
Otto-von-Guericke Universität
Magdeburg
Universitätsplatz 2
39106 Magdeburg
Germany
matthias.gundermann@vst.uni-
magdeburg.de

**Peter Heidebrecht**
Max-Planck-Institut für Dynamik
komplexer technischer Systeme
Sandtorstraße 1
39106 Magdeburg
Germany
heidebrecht@mpi-magdeburg.mpg.de

**Gerhard Huppmann**
MTU CFC Solutions GmbH
Christa-McAuliffe-Straße 1
Ottobrunn
81663 München
Germany
gerhard.huppmann@mtu-cfc.com

*Molten Carbonate Fuel Cells.* Edited by Kai Sundmacher, Achim Kienle, Hans Josef Pesch,
Joachim F. Berndt, and Gerhard Huppmann
Copyright © 2007 WILEY-VCH Verlag GmbH & Co. KGaA, Weinheim
ISBN: 978-3-527-31474-4

**Achim Kienle**
Max-Planck-Institut für Dynamik
komplexer technischer Systeme
Lehrstuhl für
Automatisierungstechnik/
Modellbildung
Otto-von-Guericke-Universität
Magdeburg
Universitätsplatz 2
Germany
kienle@mpi-magdeburg.mpg.de

**Mario Koch**
IPF Heizkraftwerks-
betriebsgesellschaft mbH
Brenneckestraße 4B
Magdeburg
Germany

**Michael Krasnyk**
Max-Planck-Institut für Dynamik
komplexer technischer Systeme
Sandtorstraße 1
39106 Magdeburg
Germany
miha@mpi-magdeburg.mpg.de

**Michael Mangold**
Max-Planck-Institut für Dynamik
komplexer technischer Systeme
Sandtorstraße 1
39106 Magdeburg
Germany

**Hans Josef Pesch**
Lehrstuhl für Ingenieurmathematik
Universität Bayreuth
Universitätsstraße 30
95440 Bayreuth
Germany
hans-josef.pesch@uni-bayreuth.de

**Joachim Rang**
Institut für Wissenschaftliches
Rechnen
Technische Universität Braunschweig
38092 Braunschweig
Germany

**Min Sheng**
Max-Planck-Institut für Dynamik
komplexer technischer Systeme
Sandtorstraße 1
39106 Magdeburg
Germany

**Kati Sternberg**
Lehrstuhl für Ingenieurmathematik
Universität Bayreuth
Universitätsstraße 30
95440 Bayreuth
Germany

**Kai Sundmacher**
Max-Planck-Insitute for Dynamics of
Complex Technical Systems
Sandtorstraße 1
39106 Magdeburg
Germany
sundmacher@mpi-magdeburg.mpg.de

# Part I
# Design and Operation

*Molten Carbonate Fuel Cells.* Edited by Kai Sundmacher, Achim Kienle, Hans Josef Pesch,
Joachim F. Berndt, and Gerhard Huppmann
Copyright © 2007 WILEY-VCH Verlag GmbH & Co. KGaA, Weinheim
ISBN: 978-3-527-31474-4

# 1
# MTU's Carbonate Fuel Cell HotModule

*Gerhard Huppmann*

## 1.1
### The Significance of Fuel Cells

Fuel cell technology and its applications are basic innovations, at best comparable to, e.g. implementations of steam engines in earlier decades of the century or the electrodynamic principle for energy conversion. For the first time in the history of energy technology fuel cells offer an alternative to thermodynamic power conversion without the efficiency limits imposed by Carnot's law, i.e. their electrical efficiency is not mainly correlating to the range of operating temperatures.

The various fuel cell technologies cover a wide scope of applications ranging from battery applications through electrical propulsion up to power stations due to their inherent modularity, high efficiency, and cleanliness. Highly reliable energy supply of small units like laptops, consumer electronics, units in spacecrafts and aircrafts, applications for supply of automotive electrical propulsion systems and auxiliary power units (APU) and as modular basic building blocks for stationary power production systems as well as combined heat and power units (CHP) or tri-generation units (combined heat, power, cooling energy) can be realised with fuel cells. The fuel cell technologies in principle differ in the utilized electrolyte giving them their names and in operating temperatures, which basically determines the ranges of possible applications and the utilization of different fuels.

The list of possible usable fuels is long and comprises
- pure hydrogen,
- gaseous or gasified hydrocarbons (natural gas, biogas, sewage gas, coal mine gas, methane containing gas mixtures, etc.),
- synthesis gases (mixtures of hydrogen and carbon monoxide)

As oxidant both pure oxygen and air are used.

Fuel cells are operating inherently clean, produce hardly emissions and offer maximum electrical efficiency.

The basic functional mechanisms of the different fuel cell types already are addressed in literature, so we have only to discuss the differences of the two temperature classes of fuel cells:

*Molten Carbonate Fuel Cells.* Edited by Kai Sundmacher, Achim Kienle, Hans Josef Pesch, Joachim F. Berndt, and Gerhard Huppmann
Copyright © 2007 WILEY-VCH Verlag GmbH & Co. KGaA, Weinheim
ISBN: 978-3-527-31474-4

- low temperature fuel cells,
  - alcaline fuel cell (AFC),
  - phosphoric acid fuel cell (PAFC),
  - proton exchange membrane fuel cell (PEM or PEMFC)
- high temperature fuel cells
  - carbonate fuel cell, often called molten carbonate fuel cell (CFC or MCFC)
  - solid oxide fuel cell, ceramic fuel cell (SOFC)

This classification of fuel cells already gives indications to their utilization and applications:

Due to their high operating temperature (approximately 600 °C up to 1000 °C) high temperature fuel cells can be equipped with facilities for internal reforming of the fuel gas and therefore are not limited to use pure hydrogen as fuel but can use gaseous or gasified hydrocarbons of all kinds. This leads to most simple system design and highest efficiency and useful high temperature heat for CHP, tri-generation and also poly-generation, which means additional use of the reaction products, e.g. utilization of $CO_2$ for fertilizing purposes or water vapour for humidification.

But due to their relatively high temperature the systems are thermally slow and consequently slow in load changes, which makes them mostly usable for stationary applications in the field of energy conversion.

Low temperature fuel cells due to their lower operating temperature have faster operational behaviour and higher dynamic in power output. Additionally they have higher energy density, meaning lower specific mass and volume. They are adaptable to the requirements of automotive propulsion, of APUs or to smaller and very small power supply units, which can never be realised by a MCFC system.

But of all existing or emerging fuel cell technologies the carbonate fuel cell (CFC) is specifically suited for stationary co-generation applications in small to medium power range (several hundred kilowatts up to several megawatts). At a temperature level of 650 °C the CFC incorporates all the advantages of high temperature fuel cells without having to cope with the problems of high temperature ceramic fuel cell manufacturing.

Triggered by good reasons MTU CFC Solutions GmbH has chosen the CFC technology in the form of the world-wide patented HotModule, which was developed for CHP, tri-generation and poly-generation in the stationary field of energy conversion. The present status is to reach maturity for a serial production of the HotModule units.

## 1.2
### Basic Statements of Power Production and Combined Heat and Power Systems

The fuel cell technology of the HotModule is a technique for today. It does not depend on realisation of the high futuristic ideas of the so-called 'hydrogen world' or 'hydrogen economy'. The HotModule uses both gaseous or gasified hydrocarbons and existing hydrogen- and/or carbon-monoxide-rich gas mixtures as fuel and as oxidant as well as for cooling medium simple air. This technology is most suitable for stationary and semi-stationary systems, which may also possibly include applications on board of ships or other big mobile systems (APU).

Putting our attention exclusively to the stationary field of power generation and neglecting for a while the requirements of mankind in the field of mobility and transportation, mankind needs only

- power,
- heat and
- cold

for basis and also for comfort of life. These energy forms we call 'consumable energies'.

In fact, all these consumable energy forms can be produced from electrical power ('all electric house'), but we should remember, that any conversion of any stored energy to electrical power is characterised by by-production of heat, which has – depending on the method – a different grade of possible utilization.

The mentioned consumable energy forms are all characterised by the fact not to be storable in necessary quantities in an economical manner. Therefore they have to be 'produced' just in time and just in right quantity. Basically, this is realised presently by the utilization of high value stored energy forms like coal, oil, natural gas and nuclear fuel. The conventional utilization of renewable energy sources (RES) is either mostly used (hydropower) or is presently emerging (wind power, biomass combustion (wood pellets), utilization of geothermal energy, etc.).

The possibilities of the applications of high temperature fuel cells, in particular of the carbonate fuel cell, to use directly gaseous or gasified hydrocarbons for conversion into power and heat are opening totally new ways for the production of consumable energy forms from energy sources till now used in small or medium quantities only.

## 1.3
## Fuels for Fuel Cells

It is one of the most important advantages of high temperature fuel cells in general and of the carbonate fuel cell in particular to have a very broad fuel flexibility.

### 1.3.1
### Fuels Containing Gaseous Hydrocarbons

The heating value of these gases is mostly based on their methane content. Further components are higher hydrocarbons, nitrogen, seldom small amounts of oxygen in fluctuating concentrations. Contaminants often are hydrogen sulphur ($H_2S$), organic sulphur (mercaptanes, thiophenes, COS), chlorine- and fluorine components, silanes and siloxanes, etc.

The most important gases of this group are:
- natural gas
- biogas from anaerobic fermentation (agricultural biogas, sewage gas, gas from industrial and municipal biogas plants)
- landfill gases
- industrial residual gases with a comparable high methane content.

## 1.3.2
### Synthesis Gases

The heating value of synthesis gases is mostly characterised by fluctuating compositions of hydrogen and carbon monoxide. They often contain nitrogen, carbon dioxide, gaseous higher hydrocarbons and mainly the same contaminants as discussed under the methane-containing gases.

The most important gases of this group are:
- Coal gas from coal gasification ('town gas'),
- Industrial residual gases with $H_2$ and CO components, e.g. furnace gases or purging gases,
- Gaseous products from thermal gasification systems of different (waste-) materials like wood, paper, cartoons, used rubber (tyres), slurry from sewage plants, organic waste material including waste from slaughter houses, hydrocarbon-containing fractions of waste, used plastics and so on.

The available techniques of gasification systems can be separated into autothermal and allothermal methods.

Autothermal gasifiers use a part of the heating value of the feed material producing the thermal energy, which is necessary for the gasification process. They need oxygen as gasification medium. As long as air is used for that, the resulting gases are characterised by a high amount of nitrogen (60% vol to 40% vol) and therefore by a low heating value. As the lower heating value of hydrogen as well as of carbon monoxide is in the range of 3 kWh/m$^3$, such gases reach a LHV of 1.5 kWh/m$^3$ or lower (for comparison only: Natural gas approximately 10 kWh/m$^3$, biogases in the range of 5 to 7 kWh/m$^3$). The alternative is to use oxygen enriched air or pure oxygen as gasification medium. Basically this is an economical decision, but technical problems should also be addressed.

Allothermal gasifiers mostly use water vapour as gasification medium, because they are heated from outside, which costs at least the same amount of fuel compared to autothermal heating. But the heating value of the produced gases is high (in the range of 3 kWh/m$^3$) due to the avoided nitrogen component. The thermal energy supply from outside can be sourced on to different fuels, namely also to un-refined product gas. Generally speaking, gases from allothermal gasification processes are much more suitable for the utilization by high temperature fuel cells.

### 1.3.3
### Group of Gasified Hydrocarbons

These include alcohols, gasoline, diesel, biodiesel, kerosene and glycerol. Within this group only alcohols, biodiesel and glycerol can be called renewable. Presently only methanol and ethanol are used for fuel cell operation.

### 1.3.4
### Secondary Fuel

At this point the term 'Secondary Fuel' in contradiction to primary energy should be defined in a more exact way: 'Secondary Fuel' or 'Secondary Energy Carrier' describe a fuel based on any material, which is already used in any way. Following this definition, secondary fuels or secondary energy carriers mostly are waste materials from natural or artificial production processes converted to gaseous or liquid fuels. Secondary fuels mostly, but not exclusively, are biogases from anaerobic fermentation and synthesis gases from thermal gasification processes. Secondary fuels are not equal to regenerative fuels, because indeed they can be produced from materials, which are originally based on fossil sources. Synthesis gas produced from waste plastics or used tyres may be an example for that. Secondary fuels mostly are biogases from anaerobic digestion and gasification of used organic material (manure, harvest residuals, wood, paper, cartoons, etc.).

In fact it is understood that the utilization of secondary fuels for conversion to consumable energies not only saves fossil sources but also reduces the atmospheric load regarding greenhouse gases, pollution and emissions being set free at the alternative utilization of primary energy sources. The reduction of political and economical dependence from crude oil and natural gas imports should be mentioned positively.

### 1.4
### Why Molten Carbonate Fuel Cells

Among all the fuel cell technologies, the molten carbonate fuel cell (CFC) is particularly suited for the stationary co-generation of electrical power and heat. This is due to its operating temperature: At approximately 650 °C, it is high enough for the electrochemical conversion processes to take place at the electrodes of the fuel cells without any precious metal catalysts. Nickel is sufficient to initiate the fuel cell reaction.

The most important reason for the utilization of high temperature fuel cells (CFC and SOFC) is their possibility to reform conventional fuel gases as well as gasified liquid or solid fuels (hydrocarbons) with the heat produced by the fuel cell itself. This internal reforming reaction, i.e. the reaction of hydrocarbons and water to form hydrogen and carbon dioxide, takes place at elevated temperatures in the presence of a catalyst inside the fuel cell block. Thanks to internal reforming, the fuel cell system greatly can be simplified. The efficiency of the system significantly is increased, because the fuel gas energy needed for reforming is saved. Additionally, carbon monoxide (CO) is a welcome fuel for the CFC in contradiction to other fuel cells, where it acts as a catalyst poison.

Another reason for the high efficiency of the CFC system is the better utilization of the heat generated in the fuel cell. High temperature exhaust heat advantageously can not only be used in industrial processes of all kinds (e.g. as process steam) but also for further power generation in downstream turbine generators, especially in larger installations. The high temperature enables the supply of heat consumers with higher temperature requests, e.g. absorption refrigerators, steam injection cooling devices, pressurised hot water production, drying processes and sterilization.

On the other hand, the operating temperature is low enough for conventional metallic materials to be used in the construction of both the cell structure and the peripheral equipment. Thus large-area fuel cells can be manufactured simply, and the peripheral equipment can be constructed cost effectively from conventional materials.

## 1.5
### The Carbonate Fuel Cell and its Function

The basic working principle of an MCFC is shown in Fig. 1.1. It basically consists of three layers, which are the porous anode and cathode electrodes and the electrolyte between these two. The metallic parts of the electrodes possess a certain electric conductivity, although they are far from being good conductors. Their thickness is only a few hundred micrometers. Nickel-based alloys are frequently used for the anode electrode, while nickel oxide is the preferred material for the cathode electrode. These materials serve as catalysts for the electrochemical reactions inside the electrode pores.

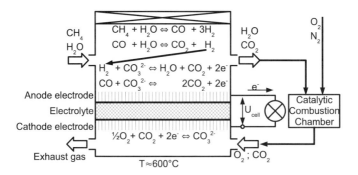

**Fig. 1.1** Work principle of the MCFC with direct internal reforming (DIR).

The electrolyte between these two is an eutectic carbonate salt mixture (e.g. $Li_2CO_3/K_2CO_3$), which is liquid at the operating temperature of about 600 °C. It is held in place by capillary forces in a porous matrix based on alloys with aluminium oxide, for example $LiAlO_2$. The layer is about as thick as the electrodes and is a fairly good conductor of carbonate ions. At the same time, this layer is a very good isolator against uncharged molecules like hydrogen or oxygen. The electrolyte is also present in the electrodes, where a part of the pores is flooded with it.

The chemical reactants of both electrodes are supplied by gas channels. The anode channel contains an additional porous catalyst for the reforming of the feed gas.

In Fig. 1.1, a mixture of methane and water is used as anode fuel gas. Upon entrance in the anode channel, these reactants undergo several reactions on the reforming catalyst, of which the methane steam reforming reaction and the water–gas shift reaction are the most important. The products of these reactions are primarily hydrogen, carbon monoxide and carbon dioxide. The concept of hydrogen production inside the anode channel is known as direct internal reforming (DIR). Although the water–gas shift reaction is slightly exothermic, the overall reforming process is endothermic:

$$CH_4 + H_2O \rightleftharpoons 3H_2 + CO \quad \text{Methane Steam Reforming}$$

$$CO + H_2O \rightleftharpoons H_2 + CO_2 \quad \text{Water–Gas Shift Reaction}$$

This endothermic process is heated by waste heat from the electrochemical reactions at the electrodes. This uptake of thermal energy leads to an increase of the heat value of the gas. In the case of a complete conversion of the reforming process, a portion of methane gas with a heating value of, say, 10 kWh will be transformed into hydrogen with a heating value of about 12 kWh. This is a significant increase in heating value using thermal waste energy. Furthermore, the endothermal character of this process possesses a kind of chemical cooling for the heat producing cell, which is important for thermal management of the system.

The reforming products, namely hydrogen and carbon monoxide, migrate into the anode electrode and react electrochemically, consuming carbonate ions from the electrolyte and producing free electrons on the electrode:

$$H_2 + CO_3^{2-} \rightleftharpoons H_2O + CO_2 + 2e^-$$
$$CO + CO_3^{2-} \rightleftharpoons 2CO_2 + 2e^-$$

The consumption of the reforming products hydrogen and carbon monoxide shifts the chemical equilibrium of the reforming process towards high conversions. Only with the direct internal reforming concept practically complete conversion of the reforming process can be obtained at a comparably low temperature of about 650 °C. Otherwise, this would require very much higher temperatures.

The anode exhaust gas consists of unreformed feed gas, reforming products and oxidation products. It is mixed with air and then fed into a catalytic combustion chamber, where all combustible species are completely oxidised. Intentionally air in stoichiometric excess is used here, so that some oxygen is left in the burner exhaust gas. This mixture is then fed into the cathode channel. Here carbon dioxide and oxygen react on the electrode producing new carbonate ions in the electrolyte and consuming electrons from the cathode electrode:

$$1/2 O_2 + CO_2 + 2e^- \rightleftharpoons CO_3^{2-}$$

The cathode exhaust gas leaves the system. The sum of the oxidation reactions with the reduction reaction corresponds to the combustion reaction of hydrogen to water or carbon monoxide to carbon dioxide, respectively. A part of their reaction enthalpy is transformed into electric energy, the rest is released as heat.

Driven by gradients in concentration and electric potential, the carbonate ions migrate through the electrolyte from the cathode towards the anode electrode. The surplus electrons on the anode are transferred to the cathode electrode, where electrons are missing, via an electric consumer, where they can perform useful electric work.

In the HotModule, the reforming process is split into three different steps, which are shown in Fig. 1.2. Outside the fuel cell, in an adiabatic external reformer (ER), short chained hydrocarbons are reformed to methane using heat from the fuel gas, which was earlier heated by the off-gas from the fuel cell, thus transforming its thermal energy into heating value. Its operating temperature is lower than the fuel cell temperature. The indirect internal reformer (IIR) is located between the cells in the cell stack. Due to the thermal coupling between the electrochemical processes in the cell and the IIR, waste heat from the cells is utilised and the reforming takes place at about cell temperature. However, no mass exchange occurs between the reforming in the IIR and the electrochemical

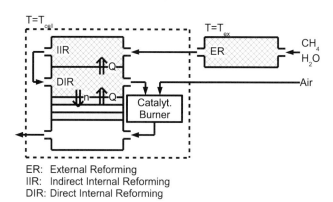

ER: External Reforming
IIR: Indirect Internal Reforming
DIR: Direct Internal Reforming

**Fig. 1.2** Reforming steps in the HotModule system.

process. This reforming step significantly increases the hydrogen content of the gas before it enters the fuel cell's anode channels. There, the direct internal reforming (DIR) continuously produces new hydrogen from the remaining methane as the electrochemical consumption of hydrogen proceeds and thereby obtains a nearly complete conversion of the reforming process.

## 1.6
## Optimisation by Integration: The HotModule Concept

The single cell is built up like a flat sandwich. Both porous nickel electrodes, the anode and the cathode wrap up the porous matrix, which is filled with electrolyte. The gas channels are supplied by corrugated current collectors, which build up the bipolar plate. The area of single cell is 0.8 m². The electrical power of one single cell is in the range of 0.8 kW. Approximately 300 cells of this kind are stacked and electrically connected serially. They are compressed by endplates and tension rods in order to ensure a good electrical and thermal contact from one to the other. Fuel gas and air form a cross flow through the stack.

The described method of the gas distribution- and gas collection facilities is called 'external manifolding'. The 'internal manifolding' is not used by MTU.

The direct current is taken from the endplates of the stack and is connected to the grid via a DC/AC converter. The excess heat is taken out from the fuel cell by the cathode airflow and can be used after using it to heat up the fuel gas and the production of steam for the humidification of the fuel gas. The temperature level for the heat utilization is around 400 °C to 450 °C. This enables the production of a high-pressure steam, for example. According to the power requirements a number of fuel cell stacks will be connected (parallel in gas flow, serial or parallel in electric flow) and completed with a fuel gas treatment system, which is adapted to the used fuel type. The inverter converts the DC to AC. The control system works fully automatically.

The fuel cell structure leads to a completely new system design. The basics are shown below.

A fuel cell power and heat co-generation plant consists of (Fig. 1.3):
- Application independent subsystems including the HotModule or some of it in a HotModule periphery, the unit for conditioning and grid connection, and the control system. All these systems are application independent and can be produced in a serial manufacturing method. They have a substantial potential for cost reduction by serial manufacturing.
- Application specific subsystems including the fuel processing unit depending on the raw gas, which shall be used, and the heat utilization facilities depending on the requirements of the customer. For cost reasons these subsystems should be built in a standardized manner.

# HotModule System

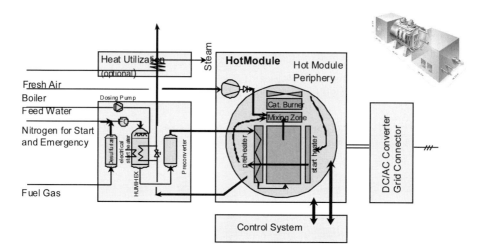

**Fig. 1.3** Subsystems of a combined heat and power plant on the HotModule basis.

The Hot Module itself is a consequent translation of the fuel cell-function principle. The design of the Hot Module was awarded the prize for innovations and future applications of natural gas sponsored by the German gas business sector on September 29, 2000.

It combines all the components of a CFC system operating at similar temperatures and pressures into a common thermally insulated vessel. A typical configuration contains the CFC stack, a catalytic burner for the anode tail gas and a cathode recycle loop including mixing-in of fresh air and anode exhaust.

The cell stack is resting horizontally on the fuel-in manifold, thus providing excellent gas sealing by gravity forces. On the top of the stack the gases leaving the anodes are mixed into the cathode recycle loop together with fresh air supplied from outside. The mixture is transported through a bed of combustion catalyst located on top of the mixing area and blown back to the cathode input by the cathode recycle blowers on the top end of the vessel. No gas piping or sealed cathode manifolds are necessary. For startup, an electrical heater is placed at the cathode input face of the stack.

## 1.7
## Manufacturing

As the mechanical components of the carbonate fuel cell consist of metallic materials they can be manufactured with conventional methods of metal sheet machining. Definition of material and application of adapted corrosion protecting layers need a lot of practical engineering and experience.

The porous electrodes and the matrix can be either produced in tape casting processes or spread to porous porter materials (Ni-foam material). The slurry for tape casting consists of metallic powder and ceramics for the electrodes and the matrix, mixed with some organic binders. A development is on the way to replace the organic binders by water soluble binders. After drying and sintering processes in drying tunnels or ovens the components will be integrated to 'unitised packages', which are roughly half cells. From these packages the fuel cell stack will be later formed.

The final conditioning of the cells will be done during the first heating up procedure in a fully integrated status. Presently, a new, different manufacturing method of producing the single cells is under development, qualification and testing. The mechanical properties of the so-called 'EuroCell' are better adapted to the horizontal position of the cell stack in the HotModule. This manufacturing process is based on new industrial available half-products. The potential of cost reduction is estimated much higher than that of the former cells. In a horizontal position every single cell is exposed to the same pressure at every position in the stack. In a vertical stack this pressure is superposed by gravity forces depending from the height of the stack. Additionally, the pressure requirements in a horizontal stack are lower, so the single cells can be defined for a lower pressurization, which saves material and costs.

## 1.8
### Advantages of the MCFC and its Utilization in Power Plants

### 1.8.1
#### Electrical Efficiency

The most important advantage of the carbonate fuel cell and its application for power plants is the high electrical efficiency with all its positive impacts to economy and the reduction of pollutant emissions as well as production of the greenhouse gas carbon dioxide. Figure 1.4 shows that the high efficiency can be reached over a very wide range of power in contrast to other technologies, which can use their advantages in smaller ranges only.

### 1.8.2
#### Modularity

Fuel cell power plants comprise a number of individual modules. The performance of such a module will be within the range of 300 kW to 1 MW. These modules will be combined to a power plant of the performance desired and equipped with a media supply unit, the necessary units for heat utilization, the DC/AC-converter and installations for the power conditioning and last but not least the

## Electrical Efficiency
## Comparison of different Power Plants

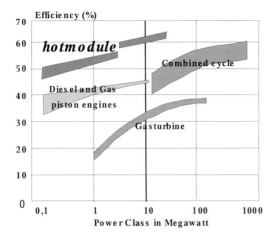

**Fig. 1.4** Comparison of electrical efficiencies.

control unit. The modular design of the fuel cell enables an industrial and serial production of the units; at users site they only have to be connected to the grid, the media supply and the utilization unit for thermal energy. This reduces construction time and cost at customer's site. Co-generation plants in the power

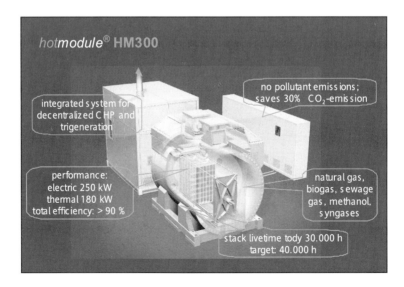

**Fig. 1.5** MTU's carbonate fuel cell HotModule.

range of 300 kW up to 30 MW can be built as well as power plants for electricity production only in the multi-MW-class using that modular design.

### 1.8.3
### Inherent Safety

Another advantage of the fuel cell units is their inherent safety feature. This is in principle due to its low energy density. No sophisticated safety and control systems are required with the consequence that cost for operating and maintaining personnel can be kept low. Based on excellent efficiency, very low level of pollutant emissions, very low risk potential combined with high reliability and availability the Fuel Cell technology offers the possibility for applications in third world countries with increasing power requirements instead of conventional types of power plants with high risk potential (nuclear power) or low environmental compatibility.

**Fig. 1.6** The first fuel cell HotModule during test at Ruhrgas, Dorsten.

### 1.8.4
### Environmentally Friendly – Pollution Free

The MCFC does not produce pollutant emissions; no $NO_x$, no sulphur components, no higher hydrocarbons, no CO. The effluent of the MCFC is not an exhaust gas but only depleted air consisting of $N_2$ (in the case of air as oxygen carrier), small amounts of $O_2$ (residual $O_2$ from excess air), large amounts of $CO_2$ (less than the amount produced by a conventional engine due to the higher electrical efficiency of the fuel cell) and big amounts of water vapour.

### 1.8.5
### Silent

Fuel cell systems are inherently silent, because they do not contain any moving parts and electrochemical reactions are – as far as they participate in fuel cells – noiseless. The only possible source of noise are blowers for gas transport, but these components can be capsuled and insulated for noise emission.

### 1.9
### History

### 1.9.1
### The European MCFC Development Consortium

The largest European program for the commercialisation of the molten carbonate fuel cell technology was carried out by the European Molten Carbonate Fuel Cell Development Consortium (ARGE MCFC). The consortium was founded in 1990 and comprised the following companies:

- MBB, Messerschmitt-Bölkow-Blohm GmbH, as predecessor of MTU Friedrichshafen GmbH (Germany), within the DaimlerChrysler Group in charge of off-road propulsion and decentralised energy systems,
- Energi E2, (Denmark), a Danish utility company, former Elkraft,
- Ruhrgas AG (Germany), a German gas company,
- RWE-Power AG (Germany), a German electrical utility company.

MBB and later MTU acted as consortium leaders. MTU shares a license and technology exchange agreement with Fuel Cell Energy Inc. (FCE), formerly known as Energy Research Corporation (ERC), Danbury, Connecticut, US.

The Consortium launched a three-phase program for the commercialisation of the MCFC technology in Europe. The overall program volume was approximately $100 million to be spent within 10 years from 1990 to 2000:

- Within the first phase the basic development of the cell and stack technology was performed with the result, that an operational lifetime of a full size stack of minimum 20,000 h can be expected with a high probability. Other targets were the development of cost-effective production processes and the basic design of layouts for natural gas and coal gas fed plants. A certain number of different plant variants were investigated and high effective and innovative plant concepts were found.
- Within the second phase cell improvement was continued in order to achieve an operational lifetime of 40,000 h. But the main issue of the second phase was the product development resulting in the concept of a highly integrated compact fuel cell power plant, the HotModule. The HotModule concept was triggered by a sustainable simplification of the MCFC power plant design and offered an important cost reduction potential.
- The third phase of the program was dominated by the design, the construction and the tests of some field test units.

During the first and second phase of the program, the ARGE MCFC spent approximately $35 million for basic technology research and development succeeding in resolving fundamental materials, corrosion, and lifetime problems associated with the MCFC technology. During this period, essential breakthroughs in the development of corrosion resistant longlife cell components have been achieved and a highly innovative system design was developed.

### 1.9.2
### Continuing of the HotModule Development at MTU CFC Solutions

The company MTU CFC Solutions GmbH was founded in early 2003 in order to collect all fuel cell development work carried out within MTU Friedrichshafen GmbH. Also in 2003 RWE Fuel Cells GmbH shared the company with a small participation. MTU CFC continued the work with the HotModule and with the EuroCell. By the end of 2004 MTU CFC together with their development and commercialisation partner FCE (Fuel Cell Energy Corp. Danbury, Connecticut, USA) had built and tested 25 plants of the HotModule type worldwide. Ten of them were erected in Europe to demonstrate their capability for combined heat and power production units. Some of them are tri-generation applications. Additionally, the ability for load-following operation as well as applications for uninterruptible power supply have been demonstrated (DeTeImmobilien, Munich). Other plants were installed at industrial environments (Michelin tyre works, Karlsruhe) or in hospitals (Bad Neustadt, Klinikum Grünstadt, Rhön-Klinikum Bad Berka).

Within 2005 and 2006 the worldwide first HotModules using biogenous gases were installed and started:

- A sewage treatment plant at Ahlen, Germany: The HotModule uses sewage gas for combined power and heat production. The heat is used partially for the sewage treatment process, the rest is fed to a district heating network. The electricity produced (230 kW) is fed to the grid.
- A municipal biological waste material treatment plant at Leonberg, Germany: Different biological waste materials collected from households, industry and public organizations is treated in a dry-fermentation process producing biogas, which is transferred to power and heat by two gas-engine modules with approximately 500 $kW_{el}$ each and a HotModule with 240 $kW_{el}$. In this plant the heat produced is completely used for a thermophilic biogas production process and a subsequent fertilizer production, the electrical energy is fed to the grid.

Also in 2006, the lifetime threshold of 30,000 operational hours was exceeded by the plant at the Magdeburg hospital. This is presently the longest operational lifetime of any fuel cell worldwide. Additionally, this fuel cell plant was used over its complete operational period as the experimental counterpart of all the investi-

# Experience:
# *hotmodule*® - Fuel Cells around the World

### Fuel Cell Energy / USA

**Coast Guard** – CHP, barracks
**Sheraton (2)** – CHP, hotels
**Ocean County College** – CHP, university
**Yale University** – CHP, university
**Zoot Enterprises (2)** – reliability for critical load
**Harrisburg Coal Mine** – power from coal mine emissions
**DFC/T (2)** – Vision 21 combined cycle
**Mercedes** – power for general load
**LADWP** – power for office headquarters
**LADWP** – CHP, WWT
**Grand Valley State University** – CHP, university
**Navy** – land-based marine diesel application
**AMP-Ohio** – utility-scale grid support at substation
**CAT Tech Centre** – grid-connected, training for engineers & dealers

### MTU / Europa

**University of Bielefeld** – CHP, university
**Rhoen-Clinic** – CHP, hospital
**RWE** – CHP, energy park
**IZAR** – power, heavy industry
**Deutsche Telecom** – DC backup, telecom
**Michelin, EnBW** – electricity/steam, tire production
**University Magdeburg** – CHP, medical clinic
**Pfalzwerke/Gruenstadt** – CHP, medical clinic
**Central-Clinic Bad Berka** – CHP, hospital
**Vattenval Europe, Bewag** –2004 CHP, bi-fuel
**FN Krefeld** – 2004 CHP district heating
**Ahlen** – 2005 CHP, sewage gas
**St. Ingbert** – 2005, CHP, industrial
**Leonberg** – 2006, CHP, biogas

### Marubeni / Japan

**Kirin Brewery** – CHP, industrial WWT
**City of Fukuoka** – CHP, municipal WWT
**Nippon Metal** – CHP, natural gas

**Fig. 1.7** HotModule field trial plants worldwide.

**Fig. 1.8** HotModule CHP at Michelin Tire Works, Karlsruhe, Germany.

**Fig. 1.9** Biogas HotModule CHP at Leonberg, Germany (small picture:
biogas storage and fermenter).

gations made by the Max Planck Institute for Dynamic of Complex Technical Systems, Magdeburg, which are presented within this publication.

## 1.10
## Possible Applications of MCFC Systems

There are to be discussed at least two types of applications of MCFC systems in the field of decentralised stationary energy supply:
- Different applications at front end, i.e. utilization of different fuels.
- Different applications at rear end, i.e. different utilizations of the products of the fuel cell system.

Both types of applications can realise additional benefits by clever integration of the fuel cell system into an overall system, where some surplus over power, heat, cooling power can be gained (components of depleted air from fuel cell system, saving of deposition costs, by-products, etc.).

### 1.10.1
### Different Applications Using Different Fuels

**Biomass Utilization**
Until now, regenerative and secondary gases are converted into electric energy in conventional CHPs with a relatively low electrical efficiency (approximately 36%) and a large quantity of low-temperature – low value – process heat production. Conventional CHPs in the performance class up to some 100 $kW_{el}$ are mostly based on gas-piston engines or diesel-injection-supported gas engines. Most of these CHP plants are operated according to the requirements of heat production with the result of short operational periods over the year (approximately 4000 h/year). Using the HotModule and its tri-generation-mode together with adapted absorption chillers, this situation can be improved to higher electrical efficiencies and better utilization of heat with the result of a full load operation all over the year. The pay back period of such an investment will decrease respectively.

The use of biomass for generation of consumable energy forms in the stationary field of power supply is one of the most important possibilities for at least a partial solution of our increasing energy problem. As biomass usable for consumable energy production is existing in two forms, namely as biomass, which is suitable to be digested within fermenters by bacteria and biomass with big amounts of ligno-cellulose, which is the main component of wood. This kind of biomass is only accessible for energy purposes by either burning or thermal gasification to synthesis gas (pyrolysis). This kind of biomass includes all sorts of wood, fresh wood, demolition wood, used wood, paper, cartoons, packing material, etc. This kind of biomass is a big component in municipal and industrial waste material. Estimations are given, that ligno-cellulose biomass represents approximately 70%

of total usable biomass. Therefore, we have to realise two different methods for utilizing biomass for consumable energy production:

- biogas utilization (biogas from digestion)
- syngas utilization (syngas from gasified biomass)

**Biogas Utilization**

By bringing the fields of biomass, (bio-)residues, anaerobic digestion and fuel cell technology together, several synergies make such applications attractive:

- Utilization of renewable energy sources (RES) in fuel cell technology – leading to a sustainable cycle by using a $CO_2$ neutral fuel. Such a fuel enhances the environmental advantage of fuel cell technology. Biogas and sewage gas is renewable energy with a very high potential for greenhouse gas reduction.
- Efficient and clean energy conversion of valuable RES: due to the nature of fuel cells, hardly any emissions are produced while converting biogas into electricity. And this is possible with high electrical efficiencies indicated above.
- High user potential for utilizing the process heat which is released from the MCFC process: due to the high temperature of the depleted air from the fuel cell system at approximately 400 °C, it is possible to use this heat in a broad variety.
- Decentralisation of the energy production is an approach for a more secure and stable energy supply. Decentralisation is one of the main advantages of RES, as these are in many cases locally available. Biogas plants are to be found usually in the decentralised agricultural sector.
- Anaerobic digestion enables a cost reduction of organic residue disposal and new income for the agricultural sector. Conventional organic waste treatment is usually strongly energy demanding, as in the case of composting. Anaerobic digestion has a higher investment cost as, e.g. composting facilities but provides the operator with energy which can be sold to the electrical grid. As organic wastes are usually co-digested in agricultural biogas plants, farmers are enabled to produce more electricity, giving them an additional income possibility.

By also involving the agricultural sector for the production of energy crops for the anaerobic digestion process it is possible to close the nutrient cycle, as the digested organic wastes are used as fertilizer on the farming land. By reducing the use of mineral fertilizers farmers contribute to the environment protection; as such fertilizers are produced with high amounts of mostly fossil energy. The digested substrate in biogas plants can substitute such fertilizers, solving in that way also the question of the disposal of these substrates.

The development of biomass digestion technology made a big step during the last decade and availability and reliability increased substantially in that period. The number of sewage treatment plants equipped with biological treatment steps also increased. The potential for biogas utilization is enormous.

### Syngas Utilization

The development of gasification systems for wood, cartoon, paper, wooden harvest residues (e.g. nut shells and other residual material) and other waste material has not yet reached industrial standard. Many systems are under development, but no one is really ready and available for application. The adaptation of the Hot-Module to different synthesis gases is under progress. Much positive experience is made with that applications in lab scale. The MCFC has been tested in many lab scale projects with success for its operability with syngas, but till now no full size HotModule is tested in an operation with syngas. Here should be mentioned an EU-project, where the adaptation of the HotModule system to different wood and waste gasification systems is under investigation (EU-Project BigPower, Project leader VTT, Finland).

### Other Fuels – Methanol

A HotModule combined heat and power production plant is under operation since September 2004 in Berlin, Germany at BEWAG facilities, which is a local

**Fig. 1.10** The methanol HotModule at BEWAG, Berlin, Germany.

utility company in Berlin (Fig. 1.10). This modified HotModule is designed for operation with methanol and all possible methanol–natural gas mixtures. As the plant has been started with natural gas for practical reasons the operation with methanol started in January 2005. Its operability with pure methanol and continuously changed mixtures of methanol and natural gas is proven meanwhile. For methanol operation the electrical efficiency reached up to 47% (see Fig. 1.4).

The principle is that the methanol is evaporated in the preheater for the fuel together with water. This principle can be used for different liquid energy carriers, e.g. ethanol, glycols, etc. The methanol in the BEWAG project is produced from plastic waste material at the facilities of SVZ in Schwarze Pumpe near Bitterfeld, Germany.

## 1.10.2
### Different Applications Using the Different Products of the MCFC System

The products of a MCFC system are:
- electrical power, originally generated in form of DC power,
- heat in form of a depleted air, originally leaving the fuel cell stack with a temperature near the operating temperature ($600\ °C$ to $650\ °C$),
- and the depleted air itself, containing the reaction products $H_2O$, $CO_2$ and $N_2$, $O_2$.

Due to the fact that most of the MCFC systems are equipped with an anode recycle, i.e. the use of the anode exhaust as part of the cathode input gas for feeding the cathode with necessary $CO_2$ and due to the implementation of a catalytic burner upstream the cathode input, the composition of the depleted air is only $N_2$, $CO_2$, and $H_2O$ vapour. No CO, no higher hydrocarbons, no sulphur due to the desulphurization prior to entrance of feed gas to fuel cell system, no fluorines, no chlorines occur in the exhaust air.

The possible applications use these products:
- stationary electrical power supply,
- heat supply,
- use of the depleted air.

### Stationary Electric Power Supply
The smallest commercial MCFC units are in the performance range of 200 to 250 $kW_{el}$ (MTU), the biggest known in the MW range (FCE). Most of these plants were installed for grid parallel operation, partially using their ability of load-following operation. But this only may be an economical solution in very particular situations, because a MCFC system is always a base load power supply unit seen from the economical point of view. This is correlated to the relatively high system cost (at least presently) combined with an excellent electrical efficiency, which only can lead under steady and full time operation conditions to a high 'economic efficiency' meaning a short pay back period of the investment costs.

Some of the realised plants were installed for a small stand-alone grid; but this is the exemption, because such grid shall be characterised by a permanent constant load, which normally cannot be granted.

Some of the realised plants were installed for DC supply in order to feed telecommunication and IT facilities via DC/DC converters.

An interesting application is the application as uninterrupted power supply (UPS), either DC or AC. Whereas a UPS normally uses three subsystems, namely the grid as base load, batteries for a short time power source and a piston engine generator for emergency power production, the fuel-cell-based UPS consists of only two subsystems, the fuel cell system for base load and the grid for emergency. Switching can be realised within microseconds.

**Heat Supply**

In most of the MCFC systems realised till now, the heat of the depleted air firstly is used for fuel treatment purposes. Fuel treatment are the functions:

- Gas clean-up (heat requirements depending on clean-up methods),
- Gas humidification. The necessary gas humidification has two reasons. One is to avoid soot formation (Boudouard-reactions), the other the availability of water vapour for the reforming step of hydrocarbons.
- Gas preheating to anode entrance temperature up to approximately 560 °C.

The enthalpy calculations result in an exit temperature of the released depleted air between 350 °C and 450 °C depending on the utilization of different feed gases. In the case of the utilization of natural gas and methane containing biogases, the average temperature is in the range of 420 °C minimum.

With such depleted air a cascade of different utilizations can be realised:

1. utilization as process heat wherever it is possible or production of high-pressure steam for any purpose,
2. production of saturated steam for any application,
3. feeding of a high temperature thermal cooling device (absorption chiller),
4. use for district heating,
5. use for low temperature heating purposes (swimming pool),
6. recondensing water from vapour and recycling as process water for humidification.

Obviously such cascade has to be adapted to customers requirements. With such heat utilization, the system becomes a CHP or a tri-generation system.

**Use of Depleted Air**

Besides the use of the enthalpy of the depleted air it can also be used as source of its chemical components. An example is the Integrated Greenhouse Supply:

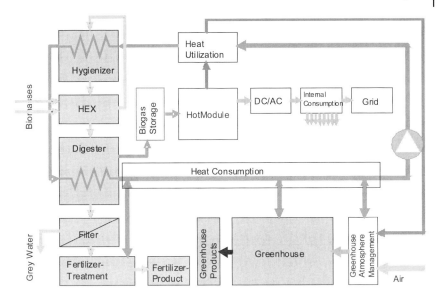

**Fig. 1.11** Integrated Greenhouse Supply with the HotModule.

Water vapour and $CO_2$ can be used in greenhouses for humidification in parallel to greenhouse heating and fertilizing. It is known that, e.g. tomatoes, cucumbers, salads and other vegetables like to have an atmosphere containing approximately 2 vol% of $CO_2$ for a accelerated growing and for increased building up of aroma compounds. The $CO_2$ fertilization is the present status of greenhouse operation; $CO_2$ normally is added from gas bottles to the greenhouse atmosphere or is produced by difficult and costly gas cleanup of exhaust of burners used for greenhouse heating or of motor based CHP systems.

A principle flow sheet of the 'Integrated Greenhouse Supply' is given in Fig. 1.11.

Due to the 'poly-generation', i.e. the utilization of the $CO_2$ additional to that of power and heat and the possibility to produce merchantable fertilizer from biogas-plant effluent, the pay-back period of the whole plant is in the range of some years only, even assumed today's high fuel cell costs. Such applications with additional benefits are representing a niche market, where already today the threshold to an economical system can be reached.

## 1.11
## Economical Impacts

High temperature CHP systems are economically suitable at investment costs of €3,000/kW, in particular, if revenues can be gained from otherwise unused waste material or by avoiding deposition costs. According coincident investigations a

Molten Carbonate Fuel Cell
## Performance, efficiencies and Costs

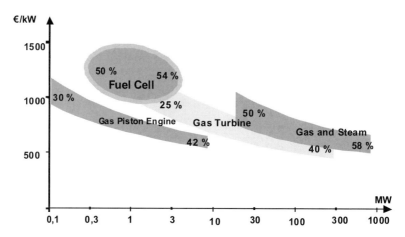

**Fig. 1.12** Comparison of specific investment costs of different stationary power plants in dependence to their typical performance class. The range of electrical efficiency is also mentioned.

big market opens at prices lower than €1,500/kW for sub-MW-plants (electrically). As distributed and decentralised installations for power production and CHP are the focus of fuel cell systems applications, the advantageous performance class is between 300 kW$_{el}$ and some megawatts, see Fig. 1.12. The figure shows the specific costs of different stationary power plants in different typical power ranges. Because all of these systems are introduced to the market, they are all cost wise mutually compatible. The target area for the HotModule is marked with a green ellipse. Its specific price can be slightly higher due to its higher efficiency and the lower expected maintenance costs. Presently the manufacturing costs of MCFC systems are for economical applications too high due to their developments status and a missing market penetration, which do not allow a serial manufacturing. This 'hen and egg' problem only can be solved by financial market entry subsidizing, which allows to sell systems at an artificial 'economic' price, thus developing the market. Besides strong simplification of the system as it is already done by the implementation of the HotModule principle the next important step in order to bring down the manufacturing costs to economic values is to build up a serial manufacturing line for the big amounts of repeating parts of the fuel cell system, which are the single cells. A serial production line for cells and their components is already under construction in MTU CFC's facilities near Munich, Germany. The startup of this production line is estimated for summer 2008, the planned capacity is between 20 MW and 50 MW plant performance per year.

# 2
# Operational Experiences

*Mario Koch, Joachim Berndt, and Matthias Gundermann*

## 2.1
## Combined Heat and Power Plant of the Company IPF in Magdeburg

The experimental work within the scope of the joint research project 'Optimized Process Control of Fuel Cell Systems with Methods of Nonlinear Dynamics' was accomplished in cooperation with the company IPF Heizkraftwerksbetriebsgesell-schaft mbH. IPF has considerable experiences concerning the construction and operation of combined heat and power plants (CHP plants) for the local energy supply and is currently increasingly engaged in the field of fuel cell technology. This medium-sized enterprise has been operating a CHP plant, which can be found near the local university hospital, for more than ten years now (see Fig. 2.1). This plant supplies the clinical institutions and an affiliated research center with heat in the form of hot water and low pressure steam as well as with cold water, electricity and emergency power respectively. For this purpose, the CHP plant possesses various conventional energy conversion plants including three gas engines, two high temperature boilers, a steam converter, two saturated steam boilers and two refrigerating machines which are able to provide a total of 2.5 MW electrical energy, 23.5 MW heat and 2 MW refrigerating capacity. The primary energy source of the power plant is natural gas. Additionally there is also extra light fuel oil available.

## 2.2
## The HotModule in Magdeburg

Besides the conventional plants the IPF has also been running a MCFC high-temperature fuel cell in their facilities in Magdeburg since October 2002. This fuel cell is a so-called HotModule with a maximum electric power output of 250 kW (see Fig. 2.2). It is the first of its kind that was installed in the newly formed German states and offers all involved research institutes the possibility to compare their theoretical conclusions directly with the measurements from a real

*Molten Carbonate Fuel Cells.* Edited by Kai Sundmacher, Achim Kienle, Hans Josef Pesch, Joachim F. Berndt, and Gerhard Huppmann
Copyright © 2007 WILEY-VCH Verlag GmbH & Co. KGaA, Weinheim
ISBN: 978-3-527-31474-4

**Fig. 2.1** The CHP (combined heat and power plant) of the IPF at the university hospital in Magdeburg, Germany.

**Fig. 2.2** The HotModule plant at the IPF in Magdeburg, Germany.

plant. Especially the neighbourship to the local University and the Max Planck Institute in Magdeburg allowed a very close cooperation during the experimental work within the scope of the validation of the process models as well as during the testing of the developed process control concepts.

The fuel cell system, which is installed at the CHP plant of the IPF, is part of a Europe-wide field trial of the manufacturer MTU CFC Solutions. During this field trial the HotModule was tested in different fields of application. In this regard, the use in hospitals makes especially high demands on the reliability and the exhaust emissions of power plants. Fuel cells are predestinated to ensure a high-quality energy supply as they do not eject harmful substances and they are furthermore noticeably quieter than most of the other systems (for example gas motors), which can only be used in clinical institutions if special noise protection measures are applied. The use of fuel cells can improve the reliability of the energy supply because in the case of a power failure they provide the required start energy for the activation of the emergency power supply.

The clean technology of the energy conversion and the relevancy for the sustainability of the energy supply were crucial for the decision of the IPF to invest in a fuel cell plant. The determining factors for the choice of the HotModule manufactured by the MTU were the electrical efficiency of the plant of up to 50%, as well as the heat recovery at exhaust air temperatures of about 400 °C, which can be used for the generation of steam and hot water. The simultaneous use of the process heat which is suitable for the heating of rooms or for the sterilisation of surgical instruments, leads to a very high overall efficiency of the plant.

For the operation of the HotModule a new building with a base area of 320 m$^2$ was built. Besides the fuel cell plant it also possesses an auditorium, which can be used for teaching activities and colloquia (see Fig. 2.3). The foundation stone of the fuel cell building was placed in April 2002 and on 29 October 2002; the HotModule was inaugurated in the presence of high-ranking politicians and representatives from the industry. The installation of the fuel cell plant and the first operating stage was funded by the German Federal Ministry of Economy and Technology and by the DaimlerChrysler AG. Moreover, all further operation of the HotModule has been financially supported within the scope of the already mentioned joint research project of the German Federal Ministry of Education and Research. But a considerable amount of the capital investments, which were connected with the construction and the operation of the HotModule, has been raised by the IPF itself.

The technical integration of the HotModule into the existing power plant could be done by using the existing natural gas and water supply. Additionally, it was necessary to arrange the supply of the auxiliary gases hydrogen, carbon dioxide and nitrogen which are essential for the start up of the fuel cell and the inertisation in the case of an emergency shutdown. For this purpose, gas bottles and pressure tanks were installed outside the building. On the electrical side, the direct current of the HotModule is transformed into 10 kV three-phase current and is afterwards supplied to the grid. As already mentioned, the waste heat utilization has always been an important aim of the installation of the HotModule.

**Fig. 2.3** The building for the Hotmodule at the IPF in Magdeburg, Germany.

Therefore, the IPF installed a counterflow waste-heat boiler where hot water of 105 °C is produced by using the exhaust air of the fuel cell plant. In turn, this hot water can be used for the supply of heat for the clinical centre or it can even be utilis in an absorption chiller for the production of cold water with a temperature of 5 °C for climatisation.

## 2.3
### Operation Experience

The HotModule had already been conditioned by the manufacturer before the final installation in Magdeburg. In this process, for example organic binders, which are necessary for the production of the cell stack, are discharged and a first oxidation of the cathode catalyser happens. Nevertheless, the startup at the facilities in Magdeburg again required a procedure of several days. In order to avoid material damages and to grant a consistent heating-up, the temperature gradient of the cell stack was limited to 5 °C to 10 °C per hour. At defined stages of the procedure, the quantities of the auxiliary gases were adjusted manually. After reaching the operating temperature the fuel gas could be supplied and the Hot-Module could be switched into the normal operation mode.

In the beginning, the operation of the fuel cell plant turned out to be difficult. Typical problems such as leakages or failures of measurement sensors had to be eliminated. However, since the beginning of its operation the HotModule has been working continuously in a fully automatic mode and in May 2006 reached the remarkable value of 30,000 operating hours. This corresponds to an availability of 95% and a new long-term record for this type of fuel cell was established. It is worth mentioning that only slight hints of degradation were determined during the entire operating time. The main reason for this is the fact that the plant was mainly operated within the medium power range.

Concerning the reliability of the HotModule, it can be stated that the fuel cell stack, which is the main part of the plant, is characterised by a very high reliability. Nearly all malfunctions which occurred during the field trial, originated from the peripheral devices. Especially the electrical inverter caused some failures of the fuel cell which could more or less be traced back to software problems. Thanks to some adjustments of the relevant programs it was possible to realize a stable operation mode of the fuel cell. Admittedly, the HotModule is currently not yet suitable for an isolated operation.

The operation of the fuel cell plant requires a daily inspection of some peripheral devices. Furthermore, continuous maintenance with a frequency of several weeks or months is necessary. The daily control contains for example the acquisition of the meter readings and the inspection of the water treatment plant. Furthermore, the cell voltages and temperatures have to be checked. The regular maintenance also includes the cleaning and the exchange of filters and filter pads, the lubricating of both recycle blowers, the calibration of the gas warning indicator as well as diverse performance checks. Thereby special attention must be paid to the water treatment as well as to the purification of the fuel gas via activated carbon filters. In order to find out if an exchange of the activated carbon is necessary, regular gas samples have to be taken and analysed. This procedure is quite time consuming and cost intensive. Therefore, another solution for the gas purification is desirable. Nevertheless, compared to similar conventional plants the maintenance effort for the HotModule is significantly lower.

In the course of the research project numerous measurements at the fuel cell were carried out in cooperation with the chair for Process Systems Engineering of the University of Magdeburg. Because of the fact that the HotModule was used in a field trial, this plant features more sensors than it would have in a later series production. The most important measurements are the cell current, the cell voltage, temperatures and flow rates. Furthermore, a gas chromatograph is installed in order to measure the gas composition at different locations of the system. Within the scope of the series of measurements, this gas chromatograph demanded special attention. Firstly, the leak tightness of the measuring section had to be guaranteed and secondly a continuous monitoring of the gas chromatograph and the surrounding conditions (for example the temperature in the gas measuring cabinet) was needed during the measurements.

An important achievement of the research project is a new model-based state estimator that tested under real operating conditions at the fuel cell plant. This new measuring and control method was developed by the project partners at the Max Planck Institute in Magdeburg. With the help of mathematical models and based on few measured data this state estimator is able to predict non-measurable values like temperatures inside the cell with a high precision (see Chapter 9). The measuring data which are an essential input for the state estimator are provided online parallel to the fuel cell operation, via special interfaces.

## 2.4
## Results and Outlook

From the point of view of a CHP plant operator, the development of fuel cells represents a promising local energy production technology with a very high electrical efficiency and minimal pollution. A further benefit is the usability of the process heat and with this the suitability for a combined heat and power generation. The success of the HotModule field trial demonstrated that the fuel cell technology is already sophisticated and reliable. However, for a successful market launch a noticeable cost reduction is necessary as the operation of such a plant is currently not yet profitable without any subsidies.

Within the scope of the HotModule field trial in Magdeburg valuable experiences were gained concerning the practical operation of this new technology. The activities in this field will, of course, be continued.

# Part II
# Model-based Process Analysis

*Molten Carbonate Fuel Cells.* Edited by Kai Sundmacher, Achim Kienle, Hans Josef Pesch,
Joachim F. Berndt, and Gerhard Huppmann
Copyright © 2007 WILEY-VCH Verlag GmbH & Co. KGaA, Weinheim
ISBN: 978-3-527-31474-4

# 3
# MCFC Reference Model

*Peter Heidebrecht, and Kai Sundmacher*

## 3.1
## Model Hierarchy

Within this work modelling serves multiple purposes: understanding the basic MCFC system behaviour, performing system analysis, predicting steady and transient states inside a MCFC, optimisation with hundreds of iterations and finally developing control strategies for steady states and dynamic scenarios. Obviously, these aims require very different mathematical models of varying detail level and complexity. In order to minimise the modelling effort it is useful to first establish a high-level reference model and from it derive all other models that may be required for any specific task. This model hierarchy, which is shown in Fig. 3.1, is the backbone of Parts B and C of this book.

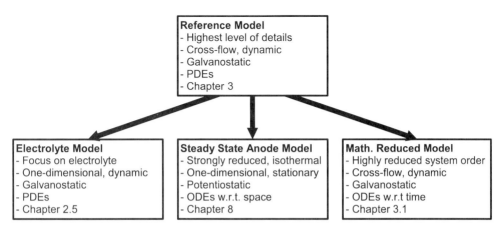

**Fig. 3.1** The hierarchy of models used in this book.

*Molten Carbonate Fuel Cells.* Edited by Kai Sundmacher, Achim Kienle, Hans Josef Pesch,
Joachim F. Berndt, and Gerhard Huppmann
Copyright © 2007 WILEY-VCH Verlag GmbH & Co. KGaA, Weinheim
ISBN: 978-3-527-31474-4

The model hierarchy is based on the so-called reference model. It describes the transient behaviour of a single cross flow MCFC. Other models are derived from the reference model. The electrolyte model focuses on the nonlinear behaviour of the electrolyte layer in the MCFC. It significantly simplifies the gas phase equations of the reference model and introduces a temperature-dependent ion conductivity. Furthermore, the spatially two-dimensional model is changed into a single spatial dimension. A different set of simplifications leads to the third model, which is the steady state anode model. The assumptions made here are strongly simplifying, turning the system into a steady state model with only one spatial dimension. Because this model is very simple and focuses on the interplay of reforming and oxidation process in the anode channel of an MCFC, it provides useful insight into the systems' principles and basic correlations. While the first two models are being derived by physically and chemically motivated simplifications, the mathematically reduced model applies a Galerkin approach to downsize the complex reference model into a small set of ordinary differential equations.

In this section, the reference model is introduced in full extent together with boundary and initial conditions and some physical interpretation of the major equations. All other models are presented in their respective Chapters 7, 8 and 9.

## 3.2
## General

The dynamic two-dimensional reference model for a single MCFC represents the highest level of modelling detail in this work. The objective is to set up a model describing transient cell behaviour suitable for system analysis, containing a large number of physical and thermodynamical details. Despite its complexity and its high numerical solution effort, the model can be applied for various purposes such as system analysis and optimisation. Other objectives such as control design or conceptual design require simpler models, which can be derived from this reference model (see Chapters 8 and 9).

The model is introduced in two steps. In this section some general assumptions and considerations are given and the most important states and parameter definitions are presented. In the subsequent section the model equations together with specific assumptions are presented and discussed in their final form. Due to their large extent no derivations of the equations are given here. For more details about this model, we refer to [1, 2]. Some other spatially distributed models of MCFC can be found in [3–8], but they shall not be discussed here in detail.

Before going into the details of the reference model, a few more words about the modelling in general are necessary: all model equations are formulated in dimensionless parameters. Dimensionless numbers such as the well-known Damkoehler number or the Reynolds number describe relations of characteristic system parameters and are frequently used, for example, in fluid mechanics, aerodynamics and chemical engineering [9]. While their application does not simplify the structure of the equations by eliminating terms or nonlinearities, this concept has certain advantages:

- The number of unknown parameters is reduced. Instead of determining the value of each unknown dimensional parameter one may assemble these into a dimensionless number which is unknown nevertheless, but then only one parameter has to be estimated or determined experimentally.
- The system description becomes more general. The results of a simulation with dimensionless parameters are valid not only for one specific cell, but for a whole class of similar systems.
- The numerical solvability is increased. If suitably defined, dimensionless numbers often have roughly the same order of magnitude, in contrast to dimensional parameters. This makes the equation system more suitable to solve for most numerical algorithms.
- Scale-up and scale-down are performed more easily.

Despite these advantages, this method also has certain drawbacks:

- For the purpose of experimental parameter validation, measurement data – which most likely possess physical units – have to be transformed into dimensionless numbers.
- Results presented in dimensionless numbers are more difficult to understand for those who are accustomed to think in dimensional parameters.

Despite these drawbacks, the mentioned advantages are dominant and thus all model equations presented in this work are completely in dimensionless form. In the appendix the reader finds a complete list of all dimensionless quantities together with their physical interpretation. Concerning the notation, the dimensionless parameters are expressed in Greek symbols whenever possible. The well-known numbers like the Damkoehler number or the Peclet number are written in their usual notation. Only if the use of a Greek symbol could possibly cause more confusion than clarity, a Latin letter is used for dimensionless quantities.

Prior to setting the model assumptions one has to define which quantities ought to be considered in the model. To fulfil its purpose the model has to accept changes in input quantities like feed gas composition, feed gas temperature and total cell current. It must deliver information about the temperatures inside the cell as well as the cell voltage, because these output quantities are of primary interest here. Nevertheless, one cannot calculate these two without knowing other states like the concentrations, flows and electric potentials inside the cell, as they are closely related to one another. Therefore, these other quantities have to be

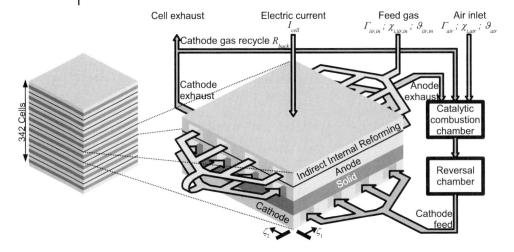

**Fig. 3.2** The model compartments are connected by mass flows. The figure also indicates the input conditions and the used coordinate system.

considered as well, so the model must contain equations for temperatures ($\vartheta$), concentrations (in terms of mole fractions, $\chi_i$), mass flows ($\gamma$ or $\Gamma$, respectively) and electric potentials ($\phi$).

Although the stack of the HotModule is a three-dimensional block consisting of 342 cells, we consider a representative cell here. This cell is an average of all stack cells. Especially at both ends of the stacks cells behave significantly different from the others due to thermal gradients, but only a few cells are affected by this. Thus it is sufficient to model only one single cell.

Figure 3.2 shows the compartments of the cell: indirect internal reforming (index: iir), anode channels ($a$), catalytic combustion chamber ($b$), reversal chamber ($m$) and cathode channels ($c$). They are connected by flows of mass and current. Also the used coordinate system is indicated. The input parameters of the model describe the amount (in terms of dimensionless molar flows, $\Gamma$), composition (in terms of mole fractions, $\chi_i$) and temperature ($\vartheta$) of the anode and air inlet, the cathode gas recycle ratio ($R_{\text{back}}$) and the total cell current ($I_{\text{cell}}$).

Due to the flat shape of the cell, assumptions are made that allow us to model the system with only two geometric dimensions. Because of the dominant horizontal or vertical flow directions in some compartments, spatial gradients occur only along one of the spatial coordinates. The general properties and assumptions of the model are as follows:

- There are no gradients in concentrations, velocities and temperatures perpendicular to the cell plane. For the three gas phases (iir, $a$, $c$) this corresponds to a plug flow profile. It also includes that there are no concentration gradients of the reactants in the electrodes and no temperature gradients in this direction within the electrode/electrolyte compartment or the bipolar plate. This simplifies the model to a spatially two-dimensional system. Please note that two different phases such as the anode gas phase and the cathode gas phase may still have different temperatures, but no temperature gradients will appear inside each of these phases.

- The cell consists of numerous parts such as two electrodes, a solid electrolyte matrix, a liquid electrolyte, channel walls, a separator or bipolar plate and gaskets, and it contains gas in the channels and inside the porous electrodes. The temperatures of some of these components, for example several of the solid parts, are almost indistinguishable and therefore they are assumed to be identical. Others show only medium heat exchange capabilities like between gas phases in the channels and the electrodes. As a result a lumped solid phase is introduced which combines the heat conductivities and heat capacities of all solid parts into one single phase. Nevertheless, its temperature is spatially distributed along the spatial coordinates. The temperatures of the gas inside the channels are calculated separately.
- The cell is part of a stack in which all cells behave alike. That leads to symmetry conditions at the boundaries to the neighbouring cells (insulated boundary).
- The HotModule is operated at nearly ambient pressure. As neither the prediction of pressure drop nor the dynamics of pressure changes are aims of the model, isobaric condition is assumed in the all gas phases.
- The catalytic combustion chamber is modelled.
- The reversal chamber between the combustion chamber outlet and the cathode inlet is modelled.

The gas phases in the indirect internal reformer, the anode and the cathode channels are all modelled under the following assumptions:

- The gas obeys to the ideal gas law.
- In each gas phase, there is a dominant gas flow direction. This is the $\zeta_1$ coordinate for the reforming unit and the anode gas, and $\zeta_2$ for the cathode gas. Gas flow perpendicular to this direction is neglected.
- Convective mass and heat transport are dominant over diffusive transport. Diffusion and heat conduction are neglected in the gas phase.
- Heat exchange between the gas phase and the solid phase is described by a linear approach. The heat exchange coefficient also includes a linearised form of the thermal radiation.

- If applicable, reforming reactions are modelled as reversible, equilibrium limited, quasi-homogeneous gas phase reactions. The methane steam reforming (indexed 'ref 1') and the water–gas shift reactions (indexed 'ref 2') are considered:

$$CH_4 + H_2O \rightleftharpoons 3H_2 + CO \tag{3.1}$$

$$CO + H_2O \rightleftharpoons H_2 + CO_2 \tag{3.2}$$

- In the case of occurrence of reforming reactions, the heat of reaction is completely transferred to the gas phase.

Concerning the electrochemical reactions, the following assumptions are made:
- Two oxidation reactions are considered to take place at the anode electrode. They are the oxidation of hydrogen (indexed ox1) and the oxidation of carbon monoxide (indexed ox2):

$$H_2 + CO_3^{2-} \rightleftharpoons H_2O + CO_2 + 2e^- \tag{3.3}$$

$$CO + CO_3^{2-} \rightleftharpoons 2CO_2 + 2e^- \tag{3.4}$$

- One reduction reaction is assumed to occur at the cathode electrode (indexed 'red'). It is noted here as an oxidation reaction, which in MCFC usually runs in the backward direction:

$$CO_3^{2-} \rightleftharpoons CO_2 + \frac{1}{2}O_2 + 2e^- \tag{3.5}$$

- The heat of reaction of any electrochemical reaction is completely transferred to the solid phase.

## 3.3
## Model Equations

This chapter presents the equations of the reference model. The order in which the equations are introduced follows the gas flow through the cell starting with the indirect internal reformer, continuing with the anode channel, the catalytic combustion chamber, the reversal chamber, the cathode channel and finally the solid phase. While some simple kinetics of heat and mass transfer are given along with the balance equations, the more complicated reaction kinetics for both the reforming and the electrochemical reactions are presented separately afterwards. All necessary thermodynamic relations are given in Section 3.3.10. For more details and an extensive discussion we refer to [1].

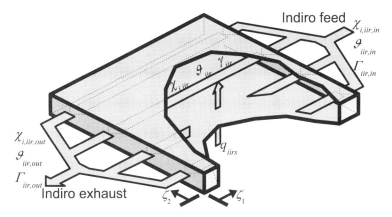

**Fig. 3.3** Anode channel with states, fluxes and boundary conditions.

## 3.3.1
**Indirect Internal Reformer**

Figure 3.3 shows the important states, inlet and outlet flows as well as the heat flux densities, which are exchanged with the solid phase. The dominant gas flow direction is in the negative $\zeta_1$ direction and it is counted positive. This influences the sign of the convective terms, which becomes clear when compared with the equations describing the anode gas phase. In this gas phase, reforming reactions occur.

The assumptions lead to the following three equations. They are the partial mass balances, the enthalpy balance in temperature form and the total mass balance, describing the mole fractions, $\chi_{i,\text{iir}}$, the gas temperatures, $\vartheta_{\text{iir}}$, and the molar flow density, $\gamma_{\text{iir}}$:

$$\frac{V_{\text{iir}}}{\vartheta_{\text{iir}}} \cdot \frac{\partial \chi_{i,\text{iir}}}{\partial \tau} = \gamma_{\text{iir}} \cdot \frac{\partial \chi_{i,\text{iir}}}{\partial \zeta_1} + \sum_{j=\text{ref}} (\nu_{i,j} - \chi_{i,\text{iir}} \bar{\nu}_j) \cdot Da_{j,\text{iir}} r_j \tag{3.6}$$

$$\frac{V_{\text{iir}}}{\vartheta_{\text{iir}}} \cdot \frac{\partial \vartheta_{\text{iir}}}{\partial \tau} = \gamma_{\text{iir}} \cdot \frac{\partial \vartheta_{\text{iir}}}{\partial \zeta_1} + \sum_{j=\text{ref}} \frac{-\Delta_R h_j^0}{\bar{c}_{p,\text{iir}}} \cdot Da_{j,\text{iir}} r_j + \frac{q_{\text{iirs}}}{\bar{c}_{p,\text{iir}}} \tag{3.7}$$

$$0 = \frac{\partial \gamma_{\text{iir}} \vartheta_{\text{iir}}}{\partial \zeta_1} + \sum_{j=\text{ref}} \frac{-\Delta_R h_j^0}{\bar{c}_{p,\text{iir}}} \cdot Da_{j,\text{iir}} r_j + \frac{q_{\text{iirs}}}{\bar{c}_{p,\text{iir}}} + \vartheta_{\text{iir}} \cdot \sum_{j=\text{ref}} \bar{\nu}_j Da_{j,\text{iir}} r_j \tag{3.8}$$

The average heat capacity of the gas mixture is calculated from

$$\bar{c}_{p,\text{iir}} = \sum_i \chi_{i,\text{iir}} c_{p,i} \tag{3.9}$$

The reaction enthalpy is temperature dependent. It is discussed in Section 3.3.10. The change of mole numbers in a reaction is described as follows:

$$\bar{v}_j = \sum_i v_{i,j} \tag{3.10}$$

The heat flux density between the solid phase and the gas is proportional to a heat transfer coefficient, the Stanton number and the temperature difference between both phases:

$$q_{\text{iirs}} = St_{\text{iirs}} \cdot (\vartheta_s - \vartheta_{\text{iir}}) \tag{3.11}$$

Details about the reaction rates are given in Section 3.3.9.

The nonlinear hyperbolic PDEs in Eqs. (3.6) and (3.7) each require one boundary and one initial condition. Equation (3.8) is a first-order ODE with respect to $\zeta_1$ requiring only one boundary condition. The boundaries are defined by the inlet gas conditions, i.e. its mole fractions, temperature and molar flow, $\chi_{i,\text{iir,in}}$, $\vartheta_{\text{iir,in}}$ and $\gamma_{\text{iir,in}}$. Note that the dimensionless molar flow density at the channel inlet equals the dimensionless total molar feed flow (Eq. (3.14)).

$$\chi_{i,\text{iir}}(\zeta_1 = 1, \zeta_2, \tau) = \chi_{i,\text{iir,in}}(\tau) \tag{3.12}$$

$$\vartheta_{\text{iir}}(\zeta_1 = 1, \zeta_2, \tau) = \vartheta_{\text{iir,in}}(\tau) \tag{3.13}$$

$$\gamma_{\text{iir}}(\zeta_1 = 1, \zeta_2, \tau) = \gamma_{\text{iir,in}}(\tau) = \Gamma_{\text{iir,in}}(\tau) \tag{3.14}$$

The initial conditions are

$$\chi_{i,\text{iir}}(\zeta_1, \zeta_2, \tau = 0) = \chi_{i,\text{iir},0}(\zeta_1, \zeta_2) \tag{3.15}$$

$$\vartheta_{\text{iir}}(\zeta_1, \zeta_2, \tau = 0) = \vartheta_{\text{iir},0}(\zeta_1, \zeta_2) \tag{3.16}$$

The terms in the balance equations can be interpreted as follows: The partial mass balances in molar form (Eq. (3.6)) describe the change of the mole fraction of any component $i$ with time due to several phenomena. The first term on the right-hand side describes the effect of convection of the gas through the channel in the negative $\zeta_1$ direction. The dimensionless molar flow density $\gamma_{\text{iir}}$ does not appear inside the derivative because the total mass balance was already inserted into this equation, eliminating $\gamma_{\text{iir}}$ in the derivative. The second and third terms describe the effect of the reforming reaction. The second term simply considers the direct effect of a component being produced or consumed in the gas phase. The third term can be interpreted to describe the dilution/concentration of reactions with changes in total mole numbers. Here, $Da_{j,\text{iir}}$ can be interpreted as the dimensionless reaction rate constant or catalyst activity, and $r_j$ is the dimensionless reaction rate.

Equation (3.7) is the dynamic enthalpy balance in the temperature form for the reformer gas. Convective enthalpy transport is considered in the first term on the right-hand side of the equation. The heat of the reforming reactions is considered in the second term, while the last term describes the convective heat exchange between the gas and solid phase.

The total molar balance (Eq. (3.8)) contains similar elements as the other two balance equations, but as a consequence of the constant pressure assumption this equation has no accumulation term. The first term can be identified as the convective term. The next two terms consider thermal expansion effects. Here again the same two phenomena are accounted for as in the temperature equation: heat of reforming reaction and heat exchange with the solid phase. The last term considers the increase of the total molar flow due to the reforming reactions.

While some of the variables mentioned here are constant system parameters (e.g. the Damkoehler number), most quantities are functions of state variables. Some of them are explicitly stated here (Eq. (3.11)), yet others may be looked up in other sections.

Before moving on to the next model compartment it is necessary to define the exhaust conditions at the end of the reformer channel ($\zeta_1 = 0$). In this model only the averaged outlet conditions are required. These are calculated by integration over the whole outlet width:

$$\Gamma_{\text{iir, out}} = \int_0^1 \gamma_{\text{iir}}(\zeta_1 = 0, \zeta_2)\, d\zeta_2 \tag{3.17}$$

$$\chi_{i,\text{iir, out}} = \int_0^1 \frac{\gamma_{\text{iir}}(\zeta_1 = 0, \zeta_2)}{\Gamma_{\text{iir, out}}} \cdot \chi_{i,\text{iir}}(\zeta_1 = 0, \zeta_2)\, d\zeta_2 \tag{3.18}$$

$$\vartheta_{\text{iir, out}} = \vartheta^0 + \int_0^1 \frac{\gamma_{\text{iir}}(\zeta_1 = 0, \zeta_2)}{\Gamma_{\text{iir, out}}} \cdot \frac{\sum_i \chi_{i,\text{iir}}(\zeta_1 = 0, \zeta_2) c_{p,i}}{\sum_i \chi_{i,\text{iir, out}} c_{p,i}}$$
$$\cdot (\vartheta_{\text{iir}}(\zeta_1 = 0, \zeta_2) - \vartheta^0)\, d\zeta_2 \tag{3.19}$$

### 3.3.2
### Anode Channel

The states, fluxes and boundary conditions of the anode channel are shown in Fig. 3.4. The dominant flow direction in this phase is in the positive $\zeta_1$ direction. The following three equations describe the important states in this phase. They are the partial mass balances, the enthalpy balance in the temperature form and the total mass balance, describing the mole fractions, $\chi_{i,a}$, the gas temperatures, $\vartheta_a$, and the molar flow density, $\gamma_a$:

$$\frac{V_a}{\vartheta_a} \cdot \frac{\partial \chi_{i,a}}{\partial \tau} = -\gamma_a \cdot \frac{\partial \chi_{i,a}}{\partial \zeta_1} + \sum_{j=\text{ref, ox}} (\nu_{i,j} - \chi_{i,a}\bar{\nu}_j) \cdot Da_j r_j \tag{3.20}$$

$$\frac{V_a}{\vartheta_a} \cdot \frac{\partial \vartheta_a}{\partial \tau} = -\gamma_a \cdot \frac{\partial \vartheta_a}{\partial \zeta_1} + \sum_{j=\text{ox}} \sum_i \nu_{i,j}^+ \cdot \frac{c_{p,i}}{\bar{c}_{p,a}} \cdot Da_j r_j \cdot (\vartheta_s - \vartheta_a)$$
$$+ \sum_{j=\text{ref}} \frac{-\Delta_R h_j^0}{\bar{c}_{p,a}} \cdot Da_j r_j + \frac{q_{as}}{\bar{c}_{p,a}} \tag{3.21}$$

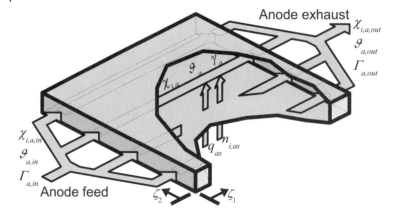

**Fig. 3.4** Anode channel with states, fluxes and boundary conditions.

$$0 = -\frac{\partial \gamma_a' \vartheta_a}{\partial \zeta_1} + \sum_{j=\text{ox}} \sum_i v_{i,j}^+ \cdot \frac{c_{p,i}}{\bar{c}_{p,a}} \cdot Da_j r_j \cdot (\vartheta_s - \vartheta_a)$$

$$+ \sum_{j=\text{ref}} \frac{-\Delta_R h_j^0}{\bar{c}_{p,a}} \cdot Da_j r_j + \frac{q_{as}}{\bar{c}_{p,a}} + \vartheta_a \cdot \sum_{j=\text{ref,ox}} \bar{v}_j Da_j r_j \qquad (3.22)$$

The average heat capacity of the gas mixture is calculated from

$$\bar{c}_{p,a} = \sum_i \chi_{i,a} c_{p,i} \qquad (3.23)$$

The reaction enthalpy is temperature dependent. It is discussed in Section 3.3.10. The change of mole numbers in a reaction is described as follows:

$$\bar{v}_j = \sum_i v_{i,j} \qquad (3.24)$$

The heat flux density between the solid phase and anode gas is proportional to a heat transfer coefficient, the Stanton number and the temperature difference between both phases:

$$q_{as} = St_{as} \cdot (\vartheta_s - \vartheta_a) \qquad (3.25)$$

In comparison to the equations for the indirect internal reforming compartment, several new terms appear here. The are related to the mass and heat effects of the electrode reactions, namely the oxidation reactions. Their contribution to the composition of the gas is taken into account in the last term in Eq. (3.20). The second term on the right-hand side of the enthalpy balance (Eq. 3.21) describes the effect caused by the reaction products leaving the electrode at solid temperature, which may be different from the anode gas temperature. For example, if the electrode is hotter than the anode gas, the oxidation products pouring out of the electrode will cause a temperature increasing effect on the anode gas. On the other hand, gas flowing from the colder anode gas into the hot electrode does not affect the gas phase temperature. Therefore, only the reaction products (denoted by $v_{i,j}^+$) have to be considered in this term. In the total mass balance (Eq. (3.22)), a similar term considers gas expansion effects due to temperature change.

Details about the reaction rates are given in Section 3.3.9.

The nonlinear hyperbolic PDEs in Eqs. (3.20) and (3.21) each require one boundary and one initial condition. Equation (3.22) is a first-order ODE with respect to $\zeta_1$ requiring only one boundary condition. The boundaries are defined by the inlet gas conditions, i.e. its mole fractions, temperature and molar flow, $\chi_{i,a,in}$, $\vartheta_{a,in}$ and $\gamma_{a,in}$. Note that the dimensionless molar flow density at the anode channel inlet equals the dimensionless total molar feed flow (Eq. (3.28)).

$$\chi_{i,a}(\zeta_1 = 0, \zeta_2, \tau) = \chi_{i,a,in}(\tau) = \chi_{i,iir,out}(\tau) \tag{3.26}$$

$$\vartheta_a(\zeta_1 = 0, \zeta_2, \tau) = \vartheta_{a,in}(\tau) = \vartheta_{iir,out}(\tau) \tag{3.27}$$

$$\gamma_a(\zeta_1 = 0, \zeta_2, \tau) = \gamma_{a,in}(\tau) = \Gamma_{a,in}(\tau) = \Gamma_{iir,out}(\tau) \tag{3.28}$$

The averaged anode outlet conditions are calculated similarly to the IIR outlet conditions:

$$\Gamma_{a,out} = \int_0^1 \gamma_a(\zeta_1 = 1, \zeta_2)\, d\zeta_2 \tag{3.29}$$

$$\chi_{i,a,out} = \int_0^1 \frac{\gamma_a(\zeta_1 = 1, \zeta_2)}{\Gamma_{a,out}} \cdot \chi_{i,a}(\zeta_1 = 1, \zeta_2)\, d\zeta_2 \tag{3.30}$$

$$\vartheta_{a,out} = \vartheta^0 + \int_0^1 \frac{\gamma_a(\zeta_1 = 1, \zeta_2)}{\Gamma_{a,out}} \cdot \frac{\sum_i \chi_{i,a}(\zeta_1 = 1, \zeta_2) c_{p,i}}{\sum_i \chi_{i,a,out} c_{p,i}}$$
$$\cdot (\vartheta_a(\zeta_1 = 1, \zeta_2) - \vartheta^0)\, d\zeta_2 \tag{3.31}$$

### 3.3.3
### Combustion Chamber

The specific assumptions for the catalytic combustion chamber are as follows:
- The combustion chamber has three inlets: the anode exhaust, the cathode recycle and the fresh air inlet (Fig. 3.5).
- The volume of the combustion chamber is neglected. Any mass and energy accumulation capacities of the burner are accounted for in the equations of the reversal chamber.
- As the combustion chamber in the HotModule is surrounded by hot compartments its heat losses are negligible.

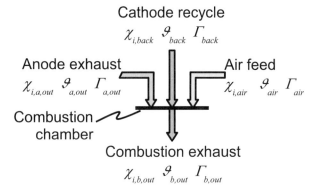

**Fig. 3.5** Catalytic combustion chamber with three inlet flows and one exhaust flow.

- The oxidation in the combustion chamber is total and instantaneous. No restrictions like limiting mass transfer, slow reaction kinetics or chemical equilibrium are considered.
- Due to a small mixing chamber in front of the combustion chamber no spatially distributed gas composition from the anode or the cathode is considered, but averaged mole fractions and temperatures are used.

The combustion equations are the total and the partial mass balances and the enthalpy balance. They are used to calculate the amount, $\Gamma_{b,\text{out}}$, composition, $\chi_{i,b,\text{out}}$, and temperature, $\vartheta_{b,\text{out}}$, of the combustion exhaust gas. The total mass balance sums the molar flows of all inlet flows modified by the change of total mole numbers due to oxidation of combustible components:

$$\Gamma_{b,\text{out}} = \Gamma_{a,\text{out}} \cdot \left(1 + \sum_j \bar{\nu}_{Cj} \cdot \chi_{j,a,\text{out}}\right)$$

$$+ \Gamma_{\text{back}} \cdot \left(1 + \sum_j \bar{\nu}_{Cj} \cdot \chi_{j,\text{back}}\right) + \Gamma_{\text{air}} \cdot \left(1 + \sum_j \bar{\nu}_{Cj} \cdot \chi_{j,\text{air}}\right) \quad (3.32)$$

The partial mass balances are set up in a similar manner:

$$\Gamma_{b,\text{out}} \cdot \chi_{i,b,\text{out}} = \Gamma_{a,\text{out}} \cdot \left(\chi_{i,a,\text{out}} + \sum_j \nu_{i,Cj} \cdot \chi_{j,a,\text{out}}\right)$$

$$+ \Gamma_{\text{back}} \cdot \left(\chi_{i,\text{back}} + \sum_j \nu_{i,Cj} \cdot \chi_{j,\text{back}}\right)$$

$$+ \Gamma_{\text{air}} \cdot \left(\chi_{i,\text{air}} + \sum_j \nu_{i,Cj} \cdot \chi_{j,\text{air}}\right) \quad (3.33)$$

The enthalpy balance in the temperature form demands that all enthalpy entering the system via the three inlet flows has to leave the system with the exhaust flow:

$$\Gamma_{b,\,out} \sum_i \chi_{i,\,b,\,out} \cdot c_{p,\,i} \cdot (\vartheta - \vartheta_{b,\,out})$$

$$= \Gamma_{a,\,out} \sum_i \chi_{i,\,a,\,out} \cdot (c_{p,\,i} \cdot (\vartheta_{a,\,out} - \vartheta^\vartheta) - \Delta_C h_i^0(\vartheta^\vartheta))$$

$$+ \Gamma_{back} \sum_i \chi_{i,\,back} \cdot (c_{p,\,i} \cdot (\vartheta_{back} - \vartheta^\vartheta) - \Delta_C h_i^0(\vartheta^\vartheta))$$

$$+ \Gamma_{air} \sum_i \chi_{i,\,air} \cdot (c_{p,\,i} \cdot (\vartheta_{air} - \vartheta^\vartheta) - \Delta_C h_i^0(\vartheta^\vartheta)) \tag{3.34}$$

Here the enthalpy of combustion, $\Delta_C h_i^0(\vartheta^\vartheta)$, is calculated from the enthalpies of formation weighted by the stoichiometric coefficients of the combustion reactions at a representative temperature:

$$\Delta_C h_i^0(\vartheta^\vartheta) = \sum_j v_{j,\,Ci} \cdot \Delta_f h_j(\vartheta^\vartheta) \tag{3.35}$$

The change in total mole numbers due to a combustion reaction is calculated from

$$\bar{v}_{Cj} = \sum_i v_{i,\,Cj} \tag{3.36}$$

The total molar balance at the burner sums up all three inlet flows and adds the total molar change of all combustion reactions. The equation can be used to calculate the amount of the burner exhaust gas. The concentrations in the burner exhaust gas are derived from partial molar balances at the burner. According to the assumption of complete combustion, no combustible components occur in the burner exhaust. The enthalpy balance is an implicit equation for the burner exhaust temperature. It can be manipulated such that the desired exhaust temperature can be calculated explicitly.

The three inlet flows have yet to be defined. The first stream comes from the anode exhaust, so its state is identical to the averaged anode exhaust (Eqs. (3.29), (3.30), and (3.31)). The conditions of the cathode gas recycle are given in Eqs. (3.56), (3.57) and (3.58) of Section 3.3.5. The conditions at the air inlet are given by the concentrations of fresh air, $\chi_{i,\,air}$, its temperature, $\vartheta_{air}$, and the air number, $\lambda_{air}$:

$$\Gamma_{air} = \Gamma_{iir,\,in} \cdot \frac{\lambda_{air}}{\chi_{O_2,\,air}} \cdot \left( \sum_j (-v_{O_2,\,Cj} \cdot \chi_{j,\,iir,\,in}) - \chi_{O_2,\,iir,\,in} \right) \tag{3.37}$$

### 3.3.4
### Reversal Chamber

A space of considerable size is located between the combustion chamber exhaust and the cathode inlet. Its sole purpose is to bridge the gap between these two compartments, so no reaction takes place here. The only thing that needs to be considered is the heat loss, which occurs across the outer walls of the containment vessel. For steady state modelling the heat loss could be incorporated into the burner equations and the volume could be neglected without loss of model precision. For a dynamic model however, this compartment has its own dynamic properties which influence the interactions between the anode exhaust and cathode inlet. The following assumptions are made for this compartment (Fig. 3.6):

- The gas inside the compartment is strongly mixed, so that inhomogeneities can be neglected. Therefore it can be modelled as an ideal continuously stirred tank reactor (CSTR).
- Heat losses across the outer vessel wall are proportional to the difference between the gas and ambient temperature.
- The feed gas of the reversal chamber is the combustion exhaust.
- The gas is moved through the channels by a pair of blowers, which are located between the combustion chamber and the reversal chamber. These impose a certain energy input on the gas in the form of heat, kinetic and compression energy. While flowing through the channels, virtually all of this energy is converted to heat by friction forces. Because the gas is assumed incompressible and kinetic energy is not considered in the gas phase equations, it is assumed that the total energy input from the blowers is induced as heat.

To obtain the concentrations, gas molar flow and temperature in this compartment, the partial and total mass balances as well as the enthalpy balance are applied:

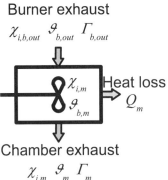

**Burner exhaust**
$$\chi_{i,b,out} \quad \vartheta_{b,out} \quad \Gamma_{b,out}$$

$\chi_{i,m}$ Heat loss
$\vartheta_{b,m}$ $Q_m$

**Chamber exhaust**
$$\chi_{i,m} \quad \vartheta_m \quad \Gamma_m$$

**Fig. 3.6** Reversal chamber with states, mass flows at inlet and outlet and heat loss.

$$\frac{V_m}{\vartheta_m} \frac{d\chi_{i,m}}{d\tau} = \Gamma_{b,\text{out}} \cdot (\chi_{i,b,\text{out}} - \chi_{i,m}) \tag{3.38}$$

$$\frac{V_m \bar{c}_{p,m}}{\vartheta_m} \frac{d\vartheta_m}{d\tau} = \Gamma_{b,\text{out}} \cdot \bar{c}_{p,b,\text{out}} \cdot (\vartheta_{b,\text{out}} - \vartheta_m) + P_{\text{blower}} - Q_m \tag{3.39}$$

$$\Gamma_m = \Gamma_{b,\text{out}} \cdot \left(1 + \frac{\bar{c}_{p,b,\text{out}}}{\bar{c}_{p,m}} \cdot \frac{\vartheta_{b,\text{out}} - \vartheta_m}{\vartheta_m}\right) + \frac{P_{\text{blower}} - Q_m}{\bar{c}_{p,m}\vartheta_m} \tag{3.40}$$

where the average heat capacities read

$$\bar{c}_{p,m} = \sum_i \chi_{i,m} \cdot c_{p,i} \tag{3.41}$$

$$\bar{c}_{p,b,\text{out}} = \sum_i \chi_{i,b,\text{out}} \cdot c_{p,i} \tag{3.42}$$

The heat loss, $Q_m$, is calculated by a heat transfer coefficient, the Stanton number, $St_m$, and the temperature difference between the reversal chamber and the surrounding air:

$$Q_m = St_m \cdot (\vartheta_m - \vartheta_u) \tag{3.43}$$

The partial mass balances (Eq. (3.38)) are a simple CSTR-type balance equation describing the effect of the inlet on the concentrations within the chamber. The steady state solution of this equation shows the equality of gas composition at inlet and outlet. The enthalpy balance in the temperature form (Eq. (3.39)) looks much alike, but in addition it considers the effect of heat losses to the environment. Contrary to the preceding equation, inlet and outlet temperatures are not equal at steady state. Both equations are ODEs of first order with respect to time and each requires an initial condition:

$$\chi_{i,m}(\tau = 0) = \chi_{i,m,0} \tag{3.44}$$

$$\vartheta_m(\tau = 0) = \vartheta_{m,0} \tag{3.45}$$

The total mass balance (Eq. (3.40)) describes the amount of gas outflow as the amount of gas inlet plus the additional flow caused by gas density changes due to temperature changes. At steady state in conjunction with the enthalpy balance it simply states the equality between mass flows at inlet and outlet.

### 3.3.5
### Cathode Channels

The equations for the gas phase are set up analogously to the anode gas phase equations. The main difference is the dominant flow direction, which in the cathode channels is along the $\zeta_2$ direction, whereas in the anode channels the gas flows perpendicularly to it, that is along the $\zeta_1$ direction:

$$\frac{V_c}{\vartheta_c} \cdot \frac{\partial \chi_{i,c}}{\partial \tau} = -\gamma_c \cdot \frac{\partial \chi_{i,c}}{\partial \zeta_2} + \sum_{j=\text{red}} (v_{i,j} - \chi_{i,c} \bar{v}_j) \cdot Da_j r_j \tag{3.46}$$

$$\frac{V_c}{\vartheta_c} \cdot \frac{\partial \vartheta_c}{\partial \tau} = -\gamma_c \cdot \frac{\partial \vartheta_c}{\partial \zeta_2} + \sum_{j=\text{red}} \sum_i \bar{v}_{i,j} \cdot \frac{c_{p,i}}{\bar{c}_{p,c}} \cdot Da_j r_j \cdot (\vartheta_s - \vartheta_c) + \frac{q_{cs}}{\bar{c}_{p,c}} \tag{3.47}$$

$$0 = -\frac{\partial \gamma_c \vartheta_c}{\partial \zeta_2} + \sum_{j=\text{red}} \sum_i \bar{v}_{i,j} \cdot \frac{c_{p,i}}{\bar{c}_{p,c}} \cdot Da_j r_j \cdot (\vartheta_s - \vartheta_c)$$

$$+ \frac{q_{cs}}{\bar{c}_{p,c}} + \vartheta_c \cdot \sum_{j=\text{red}} \bar{v}_j Da_j r_j \tag{3.48}$$

where

$$\bar{c}_{p,c} = \sum_i \chi_{i,c} c_{p,i}(\vartheta_c) \tag{3.49}$$

$$q_{cs} = St_{cs} \cdot (\vartheta_s - \vartheta_c) \tag{3.50}$$

Note that in the enthalpy balance and in the total mass balance, Eqs. (3.47) and (3.48), the heat effect of the products of the reduction reaction have to be considered. Because we noted the reduction reaction in the oxidation direction, (3.5), the reduction products are the oxidation educts, so $v_{i,j}^-$ is applied.

The boundary conditions are given by the outlet conditions of the reversal chamber:

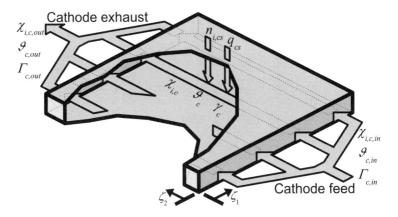

**Fig. 3.7** Cathode gas phase with states, fluxes and boundary conditions.

$$\chi_{i,c}(\zeta_1, \zeta_2 = 0, \tau) = \chi_{i,c,\text{in}}(\tau) = \chi_{i,m}(\tau) \tag{3.51}$$

$$\vartheta_c(\zeta_1, \zeta_2 = 0, \tau) = \vartheta_{c,\text{in}}(\tau) = \vartheta_m(\tau) \tag{3.52}$$

$$\gamma_c(\zeta_1, \zeta_2 = 0, \tau) = \gamma_{c,\text{in}}(\tau) = \Gamma_m(\tau) \tag{3.53}$$

Initial conditions are

$$\chi_{i,c}(\zeta_1, \zeta_2, \tau = 0) = \chi_{i,c,0}(\zeta_1, \zeta_2) \tag{3.54}$$

$$\vartheta_c(\zeta_1, \zeta_2, \tau = 0) = \vartheta_{c,0}(\zeta_1, \zeta_2) \tag{3.55}$$

The exhaust values are obtained from the following equations, which are similar to those used for the anode and reformer exhaust:

$$\Gamma_{c,\text{out}} = \int_0^1 \gamma_c(\zeta_1, \zeta_2 = 1)\, d\zeta_1 \tag{3.56}$$

$$\chi_{i,c,\text{out}} = \int_0^1 \frac{\gamma_c(\zeta_1, \zeta_2 = 1)}{\Gamma_{c,\text{out}}} \cdot \chi_{i,c}(\zeta_1, \zeta_2 = 1)\, d\zeta_1 \tag{3.57}$$

$$\vartheta_{c,\text{out}} = \vartheta^0 + \int_0^1 \frac{\gamma_c(\zeta_1, \zeta_2 = 1)}{\Gamma_{c,\text{out}}} \cdot \frac{\sum_i \chi_{i,c}(\zeta_1, \zeta_2 = 1) c_{p,i}}{\sum_i \chi_{i,c,\text{out}} c_{p,i}}$$
$$\cdot (\vartheta_c(\zeta_1, \zeta_2 = 1) - \vartheta^0)\, d\zeta_1 \tag{3.58}$$

A part of the cathode exhaust gas is recycled back into the catalytic combustion chamber, while the rest leaves the HotModule. Composition and temperature of the recycle flow is identical to the outlet, and the amount of the flow is described by a recycle ratio, $R_{\text{back}}$:

$$\chi_{i,c,\text{out}} = \chi_{i,\text{exhaust}} = \chi_{i,\text{back}} \tag{3.59}$$

$$\vartheta_{c,\text{out}} = \vartheta_{\text{exhaust}} = \vartheta_{\text{back}} \tag{3.60}$$

$$\Gamma_{\text{back}} = R_{\text{back}} \cdot \Gamma_{c,\text{out}} \tag{3.61}$$

$$\Gamma_{\text{exhaust}} = (1 - R_{\text{back}}) \cdot \Gamma_{c,\text{out}} \tag{3.62}$$

## 3.3.6
### Electrode Pores

In gas diffusion electrodes, which are applied in fuel cells, electrochemical reactions are strongly coupled with transport phenomena. These include several steps like gas film and pore diffusion, phase transition into the electrolyte and diffusion through the electrolyte [10]. For many fuel cells, these transport processes into and out of the porous electrode are the limiting processes concerning current density, so they should be taken into account in a fuel cell model. On the other hand, these coupled processes are not the only dominant phenomena in MCFC and the implementation of a detailed electrode model into this spatially distributed fuel cell model would require unacceptable numerical effort. Therefore, a simplified electrode model is applied which allows us to include mass transport resistance into the model. It is set up according to the following assumptions:

- Instead of a spatially distributed concentration profile along the pore depth, an integral balance is used to calculate a representative partial pressure for each component inside the pore.
- The gas volume inside the pores is considered negligible compared to the gas volume in the gas channels. As a consequence, the mass balances in the electrode pores are assumed to be at quasi steady state.
- The mass flux of a specific component is proportional to the difference in its partial pressure in the pore, $\varphi_{i,\,as/cs}$, and in the gas phase, $\chi_{i,\,as/cs}$. The proportionality factor, $D_{i,\,as}$ or $D_{i,\,cs}$, respectively, includes all transport effects that might occur between the reaction zone in the pore and the gas phase.
- Mass transport along the $\zeta_1$ and $\zeta_2$ coordinates is negligible.
- Contrary to all other compartments the pressure is not constant inside the pores. This assumption is necessary to enable nonzero total flux densities.
- The assumption of the ideal gas law is also valid in the electrode pores.
- Due to the high inner surface in the porous electrodes, heat exchange between the gas inside the pores and the solid parts of the electrode is very intense. It is assumed that the gas inside the electrode pores has the same temperature as the solid surrounding it. Consequently, no separate enthalpy balance is required for the gas inside the pores.
- The solid temperature changes more slowly than the partial pressures in the pores. Thus effects due to temperature-dependent gas density changes are neglected.

These assumptions lead to the following steady state partial mass balance equations: They state that the component mass source due to the electrochemical reaction is equal to the mass transport into or out of the electrode pore.

$$0 = \sum_{j=\mathrm{ox}} v_{i,j} Da_j \cdot r_j - D_{i,\,\mathrm{as}} \cdot (\varphi_{i,\,\mathrm{as}} - \chi_{i,\,\mathrm{as}}) \tag{3.63}$$

$$0 = \sum_{j=\mathrm{red}} v_{i,j} Da_j \cdot r_j - D_{i,\,\mathrm{cs}} \cdot (\varphi_{i,\,\mathrm{cs}} - \chi_{i,\,\mathrm{cs}}) \tag{3.64}$$

These equations are used to calculate the partial pressure of each component inside the pores, $\varphi_{i,\,as/cs}$. Note that the reaction rates of all electrochemical reactions,

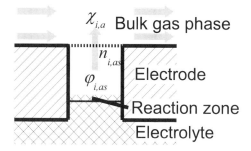

**Fig. 3.8** Anode pore with component partial pressures and mass exchange flux density between the bulk gas phase and gas phase in the pore.

$r_j$, $j \in \{ox, red\}$, depend on the partial pressures in the respective electrode. Together with the explicit reaction kinetics in Section 3.3.9, these are a nonlinear equations with respect to $\varphi_i$.

### 3.3.7
### Solid Phase

The solid phase sums up all solid, that is non-moving phases inside the MCFC. As mentioned in the general assumptions, these include the channel walls, separating plates, the electrodes and the electrolyte. The only state of interest here is the spatially distributed temperature and it is assumed to be identical in all parts of which the solid phase is comprised. Figure 3.9 shows the state and the heat fluxes of this phase. As indicated, heat exchange between solid and gas phases as well as heat conduction inside the cell plane is considered. In addition, the

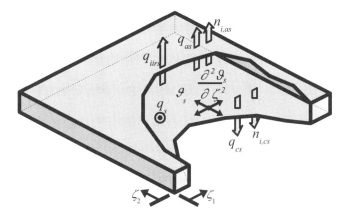

**Fig. 3.9** Solid phase states, mass and heat exchange flows with gas phases, heat conduction and heat source.

enthalpy flux due to mass transport is also taken into account. The enthalpy balance is set up under the following assumptions:

- Gas from both gas phases is brought to solid temperature at entering the electrode, influencing the solid temperature.
- The heat capacity of the gas inside the electrode pores is negligible compared to the heat capacity of the rest of the solid compartments.
- All three electrochemical reactions are taken into account. Their heat of reaction is fully attributed to the solid phase.
- The ion transport through the electrolyte induces heat in the solid phase.
- Heat losses at the outer boundary of the solid phase are negligible.

With this we obtain an enthalpy balance in the temperature form:

$$c_{p,s} \frac{\partial \vartheta_s}{\partial \tau} = \frac{l_2}{Pe_s} \frac{\partial^2 \vartheta_s}{\partial \zeta_1^2} + \frac{1}{Pe_s l_2} \frac{\partial^2 \vartheta_s}{\partial \zeta_2^2} + \sum_{j=\text{ox}} \sum_i (-v_{i,j}^-) \cdot c_{p,i} \cdot Da_j r_j \cdot (\vartheta_a - \vartheta_s)$$

$$+ \sum_{j=\text{red}} \sum_i (-v_{i,j}^+) \cdot c_{p,i} \cdot Da_j r_j \cdot (\vartheta_c - \vartheta_s)$$

$$- q_{as} - q_{cs} - q_{iirs} + q_s \tag{3.65}$$

The first two terms on the right-hand side are the heat conduction terms along both spatial coordinates. The next two terms describe the temperature effect of components flowing into the electrodes, where they are being heated up or cooled down to solid temperature. The next three terms contain the heat exchange density between the anode, cathode and reformer gas phases with the solid phase. The last term, $q_s$, is the heat source density due to electrochemical reactions and ion conduction in the electrolyte:

$$q_s = \sum_{j=\text{ox}} (-\Delta_R h_j^0(\vartheta_s) + n_j(\phi_a^S - \phi_a^L)) Da_j r_j$$

$$+ \sum_{j=\text{red}} (-\Delta_R h_j^0(\vartheta_s) + n_j(\phi_c^S - \phi_c^L)) Da_j r_j$$

$$+ (\phi_a^L - \phi_c^L) \cdot i_e \cdot \frac{1}{F} \tag{3.66}$$

In this heat source, the first two terms describe the heat of the anodic and the cathodic electrochemical reactions. The heat induced by these reactions is identical to the temperature-dependent reaction enthalpy, $\Delta_R h_j^0$, diminished by the generated electrical energy, which is the number of transferred electrons, $n_j$, multiplied by the electric potential difference at the respective electrode, for example

$\phi_a^S - \phi_a^L$. The last term multiplies the ionic current density through the electrolyte with the potential difference across the electrolyte layer to obtain the heat released due to ion migration.

The heat exchange densities are proportional to the corresponding temperature difference multiplied by a heat transfer coefficient, the Stanton number, $St$:

$$q_{as} = St_{as} \cdot (\vartheta_s - \vartheta_a) \tag{3.67}$$

$$q_{cs} = St_{cs} \cdot (\vartheta_s - \vartheta_c) \tag{3.68}$$

$$q_{iirs} = St_{iirs} \cdot (\vartheta_s - \vartheta_{iir}) \tag{3.69}$$

The boundary and initial conditions are given by the insulation condition at the boundary and the starting temperature:

$$\left. \frac{\partial \vartheta_s}{\partial \zeta} \right|_{\delta\zeta} = 0 \tag{3.70}$$

$$\vartheta_s(\zeta_1, \zeta_2, \tau = 0) = \vartheta_{s,0}(\zeta_1, \zeta_2) \tag{3.71}$$

### 3.3.8
### Electric Potential

The description of the spatially distributed electric potential in a fuel cell is essential to the behaviour of the whole model. The approach used here is based on the spatially one-dimensional version of Poisson's law and discrete charge layers. The equations are set up according to the following assumptions (Fig. 3.10):

- The electrodes combined with the current collectors are perfect electric conductors. This is equivalent to the assumption of spatially constant electric potentials at each electrode.

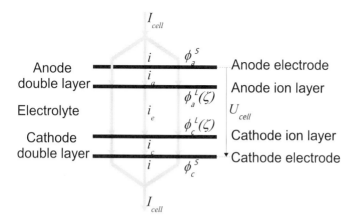

**Fig. 3.10** States and charge fluxes of the electric potential field model.

- Because only potential differences are of importance, the electric potential at the anode electrode is arbitrarily set to zero.
- The charged double layers at both electrodes are modelled as flat planes.
- Due to the extremely thin electrolyte layer, carbonate ions are assumed to be transported through the electrolyte only in a direction orthogonal to the cell plane. No ion flux is possible along the $\zeta_1$ and $\zeta_2$ coordinates.
- The charge balances at the double layers are considered transient.

The anode electrode potential is spatially constant and arbitrarily set to zero:

$$\phi_a^S = 0 \tag{3.72}$$

The spatially distributed potentials in the electrolyte in the anode and cathode double layer change depending on the electric current density flowing into the cell, $i$, and the current density produced by the electrochemical reactions, $i_a$, or by the ionic current density through the electrolyte, $i_e$:

$$\frac{\partial \phi_a^L}{\partial \tau} = -\frac{1}{c_a} \cdot (i - i_a) \tag{3.73}$$

$$\frac{\partial \phi_c^L}{\partial \tau} = -\frac{1}{c_a} \cdot (i - i_a) - \frac{1}{c_e} \cdot (i - i_e) \tag{3.74}$$

The cathode electrode potential, which is a concentrated state, depends on the differences of the overall currents produced by the electrochemical reactions respectively the overall current through the electrolyte from the given total cell current:

$$\frac{\partial \phi_c^S}{\partial \tau} = \frac{I_a - I_{\text{cell}}}{c_a} + \frac{I_e - I_{\text{cell}}}{c_e} + \frac{I_c - I_{\text{cell}}}{c_c} \tag{3.75}$$

where the total currents are the integral of the current densities:

$$I_a = \int_0^1 \int_0^1 i_a \, d\zeta_1 \, d\zeta_2 \quad I_e = \int_0^1 \int_0^1 i_e \, d\zeta_1 \, d\zeta_2 \quad I_c = \int_0^1 \int_0^1 i_c \, d\zeta_1 \, d\zeta_2 \tag{3.76}$$

The electric current is distributed across the cell according to

$$i = \left( \frac{1}{c_a} + \frac{1}{c_e} + \frac{1}{c_e} \right)^{-1} \cdot \left( \frac{i_a}{c_a} + \frac{i_e}{c_e} + \frac{i_c}{c_c} - \frac{I_a}{c_a} - \frac{I_e}{c_e} - \frac{I_c}{c_c} \right) + I_{\text{cell}} \tag{3.77}$$

Note that in dimensionless notation, current density and total current are both dimensionless and the electrode area of the cell is unity, so both can be substracted or added without conflicting units. In these equations, the anodic and cathodic current densities are calculated from Faraday's law:

$$i_a(\phi_a^L) = \sum_{j=\text{ox}} n_j F Da_j r_j(\phi_a^L) \tag{3.78}$$

$$i_c(\phi_c^L, \phi_c^S) = -\sum_{j=\text{red}} n_j F Da_j r_j(\phi_c^L, \phi_c^S) \tag{3.79}$$

The ionic current density obeys Ohm's law:

$$i_e(\phi_a^L, \phi_c^L) = \kappa_e \cdot (\phi_a^L - \phi_c^L) \tag{3.80}$$

Although it is necessary to solve Eqs. (3.73), (3.74), (3.75) and (3.77) together with the other explicit algebraic equations to simulate the complete electric potential, it actually might be just one state that is really interesting, which is the cell voltage. It is the difference of both electrode potentials:

$$U_{\text{cell}} = \phi_c^S - \phi_a^S \tag{3.81}$$

With $\phi_a^S$ equal to zero, this equation could also be used to substitute $\phi_c^S$ in Eq. (3.75).

In addition, the ODEs (Eqs. (3.73)–(3.75)) require initial conditions:

$$\phi_a^L(\zeta_1, \zeta_2, \tau = 0) = \phi_{a,0}^L(\zeta_1, \zeta_2) \tag{3.82}$$

$$\phi_c^L(\zeta_1, \zeta_2, \tau = 0) = \phi_{c,0}^L(\zeta_1, \zeta_2) \tag{3.83}$$

$$\phi_c^S(\tau = 0) = \phi_{c,0}^S \tag{3.84}$$

For the numerical solution of the model the potential field equations prove to be a critical point. The set of ordinary differential and algebraic equations (DAE) used here has a differential index of 1, so it can be solved by most numerical integrators. Nevertheless, the integration algorithm should be able to treat stiff systems, for the time constants of these ODEs, the double layer capacitances $c_a$, $c_e$ and $c_c$, are several orders of magnitude smaller than those of the gas phase balances.

### 3.3.9
### Reaction Kinetics

Two different types or reactions need to be considered in this model: chemical reactions, which occur in the reforming process both in the (indirect internal) reformer and in the anode channel, and electrochemical reactions, which take place at the anode and cathode electrode, respectively. In the scientific literature, numerous kinetic approaches are available for the steam reforming process. In

MCFC, this process usually runs close to its thermodynamic equilibrium, so the kinetics have to reflect the equilibrium correctly rather than describe a specific reaction mechanism. Consequently, the reforming reactions are described by a power law combined with an Arrhenius term to incorporate temperature effects. The reaction rate of the methane steam reforming reaction reads

$$r_{ref1} = \exp\left(Arr_{ref1} \cdot \left(\frac{1}{\vartheta^0} - \frac{1}{\vartheta_a}\right)\right) \cdot \left(\chi_{CH_4,a}\chi_{H_2O,a} - \frac{\chi_{CO,a}\chi_{H_2,a}^3}{K_{ref1}(\vartheta_a)}\right) \tag{3.85}$$

The water gas shift reaction rate is described by

$$r_{ref2} = \exp\left(Arr_{ref2} \cdot \left(\frac{1}{\vartheta^0} - \frac{1}{\vartheta_a}\right)\right) \cdot \left(\chi_{CO,a}\chi_{H_2O,a} - \frac{\chi_{CO_2,a}\chi_{H_2,a}}{K_{ref2}(\vartheta_a)}\right) \tag{3.86}$$

The equilibrium constants are calculated according to

$$K_{j=ref}(\vartheta_a) = \exp\left(-\frac{\Delta_R g_j^0(\vartheta_a)}{\vartheta_a}\right) \tag{3.87}$$

with the standard Gibbs enthalpy of reaction, $\Delta_R g_j^0$, described in Section 3.3.10.

The reaction rate expressions for the electrochemical reactions are set up under the following assumptions:

- Reversible Butler–Volmer kinetics in combination with an Arrhenius law are applied.
- The partial pressures of the gases inside the pores are the relevant concentration measures.
- Because mass transport is already considered in the pore model, they are not considered in the reaction kinetics.
- Carbonate ion concentration in the electrolyte is assumed to be constant and identical at both electrodes.

Little is known in the literature about the anodic reaction kinetics in MCFC. Fortunately, the electrochemical oxidation is fast compared to the cathodic reduction, so it is not a dominant process in today's MCFC. In this model, the orders of reaction at the anode correspond to the stoichiometric coefficient of the respective reaction.

$$r_{ox1} = \exp\left(Arr_{ox1} \cdot \left(\frac{1}{\vartheta^0} - \frac{1}{\vartheta_s}\right)\right)$$
$$\cdot \left[\varphi_{H_2,ac} \cdot \exp\left(\alpha_{ox1,+} \cdot n_{ox1}\frac{(\phi_a^S - \phi_a^L) - \Delta\phi_{ox1,0}(\vartheta_s)}{\vartheta_s}\right) - \varphi_{H_2O,ac}\varphi_{CO_2,ac}\right.$$
$$\left. \cdot \exp\left(-(1 - \alpha_{ox1,+}) \cdot n_{ox1}\frac{(\phi_a^S - \phi_a^L) - \Delta\phi_{ox1,0}(\vartheta_s)}{\vartheta_s}\right)\right] \tag{3.88}$$

$$r_{ox2} = \exp\left(Arr_{ox2} \cdot \left(\frac{1}{\vartheta^\theta} - \frac{1}{\vartheta_s}\right)\right)$$

$$\cdot \left[\varphi_{CO,ac} \cdot \exp\left(\alpha_{ox2,+} \cdot n_{ox2} \frac{(\phi_a^S - \phi_a^L) - \Delta\phi_{ox2,0}(\vartheta_s)}{\vartheta_s}\right) - \varphi_{CO_2,ac}^2\right.$$

$$\left.\cdot \exp\left(-(1 - \alpha_{ox2,+}) \cdot n_{ox2} \frac{(\phi_a^S - \phi_a^L) - \Delta\phi_{ox2,0}(\vartheta_s)}{\vartheta_s}\right)\right] \tag{3.89}$$

The standard open circuit voltage is only dependent on temperature,

$$\Delta\phi_{j,0}(\vartheta_s) = \frac{\Delta_{RG}g_j^0(\vartheta_s)}{n_j} \tag{3.90}$$

In today's MCFC, the cathodic reduction is the limiting process and therefore it deserves special attention. The exact mechanism of this reaction is still under discussion, see for example Prins–Jansen [11, 12] and others [13]. Bednarz [14] has conducted kinetic experiments with cathode electrodes from the HotModule and proposes a mechanism which is similar to the superoxide reaction mechanism found in the literature. In this model, we use the formulation of the reaction rate for the superoxide mechanism from Prins–Jansen. Some adaptations are made to introduce the temperature-dependent equilibrium voltage. Finally, the reduction reaction rate is described by

$$r_{red} = \exp\left(Arr_{red} \cdot \left(\frac{1}{\vartheta^0} - \frac{1}{\vartheta_s}\right)\right)$$

$$\cdot \left[\varphi_{CO_2,cc}^{-2} \cdot \exp\left(2.5 \cdot \frac{(\phi_c^S - \phi_c^L) - \Delta\phi_{red,0}(\vartheta_s)}{\vartheta_s}\right)\right.$$

$$\left. - \varphi_{O_2,cc}^{3/4} \cdot \varphi_{CO_2,cc}^{-1/2} \cdot \exp\left(-0.5 \frac{(\phi_c^S - \phi_c^L) - \Delta\phi_{red,0}(\vartheta_s)}{\vartheta_s}\right)\right] \tag{3.91}$$

The expressions for the electrochemical reaction rates and the electric cell potentials used in this model deviate from those expressions often found in the literature. Explicit, spatially distributed potential differences at the anode and cathode double layers are used and related to the equilibrium voltage under hypothetical standard conditions (all partial pressures $\varphi_i = 1$). If their contribution to the cell voltage is of interest, the concentration and activation overpotentials can be recalculated from this data afterwards, but they are not required to solve this model.

## 3.3.10
## Thermodynamics

Thermodynamic relations are used to calculate temperature-dependent reaction enthalpies and free (Gibbs) enthalpies of reactions. Also the definition of the electric cell power and the electric cell efficiency is given here.

The enthalpy of formation of each species $i$ is temperature dependent. For the sake of simplicity, this dependence is linearised at a representative cell temperature, $\vartheta^\theta$:

$$\Delta_f h_i^0(\vartheta) = \Delta_f h_i^0(\vartheta^\theta) + \int_{\vartheta^\theta}^{\vartheta} c_{p,i}(\vartheta)\, d\vartheta$$

$$\approx \Delta_f h_i^0(\vartheta^\theta) + c_{p,i}(\vartheta^\theta) \cdot (\vartheta - \vartheta^\theta) \tag{3.92}$$

The enthalpy of reaction $j$ at any temperature is the sum of the enthalpy or entropy of formation weighted with the stoichiometric factors of the reaction under consideration:

$$\Delta_R h_j^0(\vartheta) = \sum_i v_{i,j} \Delta_f h_i^0(\vartheta)$$

$$= \Delta_R h_j^0(\vartheta^\theta) + \Delta_R c_{p,j}^0(\vartheta^\theta) \cdot (\vartheta - \vartheta^\theta) \tag{3.93}$$

where

$$\Delta_R c_{p,j}^0 = \sum_i v_{i,j} c_{p,i} \tag{3.94}$$

The entropy of formation of any component $i$ is temperature dependent:

$$s_{f,i}^0(\vartheta) = s_{f,i}^0(\vartheta^\theta) + \int_{\vartheta^\theta}^{\vartheta} \frac{c_{p,i}(\vartheta)}{\vartheta}\, d\vartheta$$

$$\approx s_{f,i}^0(\vartheta^\theta) + c_{p,i}(\vartheta^\theta) \cdot \ln \frac{\vartheta}{\vartheta^\theta} \tag{3.95}$$

The entropy of reaction $j$ is calculated as

$$\Delta_R s_j^0(\vartheta) = \sum_i v_{i,j} s_{f,i}^0(\vartheta)$$

$$= \Delta_R s_j^0(\vartheta^\theta) + \Delta_R c_{p,j}^0(\vartheta^\theta) \cdot \ln \frac{\vartheta}{\vartheta^\theta} \tag{3.96}$$

The free (Gibbs) enthalpy of reaction is composed of the reaction enthalpy and the reaction entropy and depends of the temperature in a nonlinear fashion. Again, to simplify this, the relation is linearised. This is achieved by neglecting the temperature dependence of both the reaction enthalpy and the reaction entropy. The error of this simplification is small unless temperature deviates strongly from the standard temperature, $\vartheta^\theta$. This is mainly because the most important temperature dependence is the factor in front of the reaction entropy, which is still considered in the approximation:

$$\Delta_{R}g_{j}^{0}(\vartheta) = \Delta_{R}h_{j}^{0}(\vartheta) - \vartheta \cdot \Delta_{R}s_{j}^{0}(\vartheta)$$
$$\approx \Delta_{R}h_{j}^{0}(\vartheta^{0}) - \vartheta \cdot \Delta_{R}s_{j}^{0}(\vartheta^{0}) \tag{3.97}$$

## 3.4
## Summary

A mathematical model for a representative single MCFC cell with indirect and direct internal reforming is presented. Strictly based on dynamic balances of mass, energy and charges, the model describes changes in gas compositions, mass flux densities, temperatures, current densities and electric potential in a spatially two-dimensional cross-flow cell. All equations are formulated in terms of dimensionless expressions, so the model is valid for a whole class systems with identical dimensionless numbers. All parameters have a direct physical meaning, so the model is fully physically interpretable. The mathematical form is a system of 24 hyperbolic and 1 parabolic partial differential equations combined with several ordinary differential equations with respect to spatial coordinates or time and a large number of algebraic equations.

The reference model can be applied for different purposes, some of which are demonstrated in this book. In Chapter 4, the system is analysed from a mathematical point of view. In Chapter 5 unknown model parameters are identified using measurement information from an industrial scale MCFC, the Hotmodule. The validated model can be used to improve the understanding of the system behaviour or to predict system responses under transient boundary conditions, as is demonstrated in Chapter 6. In Chapters 10 and 11, the model is used to optimise operating conditions under steady state and transient conditions as well as design parameters of the MCFC.

The reference model is a rather complex equation system, which is too complex for some objectives. Based on the reference model, simplified models can be derived which can be tailored for specific tasks. Chapters 7 and 8 demonstrate a mathematical and a physical model reduction approach, respectively, and the application of the resulting equation systems for nonlinear system analysis and conceptual system design.

### Bibliography

1 Heidebrecht, P., *Modelling, Analysis and Optimisation of a Molten Carbonate Fuel Cell with Direct Internal Reforming (DIR-MCFC)*, VDI Fortschritt-Berichte, Reihe 3, Nr. 826, VDI-Verlag, Düsseldorf, 2005.

2 Heidebrecht, P., Sundmacher, K., Dynamic model of a cross-flow molten carbonate fuel cell with direct internal reforming (DIR-MCFC), *Journal of the Electrochemical Society* 152(1), 2005, A2217–A2228.

3 Wolf, T.L., Wilemski, G., Molten carbonate fuel cell performance model, *Journal of The Electrochemical Society: Electrochemical Science and Technology* 130(1), 1983, 48–55.

4 Yoshiba, F., Ono, N., Izaki, Y., Watanabe, T., Abe, T., Numerical analyses of the internal conditions of a molten carbonate fuel cell stack: comparison of stack performances for various gas flow types, *Journal of Power Sources* 71, 1998, 328–336.

5 Bosio, B., Costamagna, P., Parodi, F., Modeling and experimentation of molten carbonate fuel cell reactors in a scale-up process, *Chemical Engineering Science* 54, 1999, 2907–2916.

6 Koh, J.-H., Kang, B.S., Lim, H.C., Analysis of temperature and pressure fields in molten carbonate fuel cell stacks, *AIChE Journal* 47(9), 2001, 1941–1956.

7 Park, H.-K., Lee, Y.-R., Kim, M.-H., Chung, G.-Y., Nam, S.-W., Hong, S.-A., Lim, T.-H., Lim, H.-C., Studies of the effects of the reformer in an internal-reforming molten carbonate fuel cell by mathematical modeling, *Journal of Power Sources* 104, 2002, 140–147.

8 Lukas, M.D., Lee, K.Y., Ghezel-Ayagh, H., Development of a stack simulation model for control study on direct reforming molten carbonate fuel cell power plant, *IEEE Transactions on Energy Conversion*, 14(4), 1999, 1651–1657.

9 Zlokarnik, M., *Dimensional Analysis and Scale-up in Chemical Engineering*, Springer-Verlag, Heidelberg, 1991.

10 Sundmacher, K., *Reaktionstechnische Grundlagen der elektrochemischen Absorption mit Gasdiffusionselektroden*, VDI Fortschritt-Berichte, Reihe 3, Nr. 564, VDI-Verlag, Düsseldorf, 1998.

11 Prins-Jansen, J.A., Hemmes, K., de Wit, J.H.W., An extensive treatment of the agglomerate model for porous electrodes in molten carbonate fuel cells: I. Qualitative analysis of the steady-state model, *Electrochimica Acta* 85, 1997, 3585–3600.

12 Prins-Jansen, J.A., Hemmes, K., de Wit, J.H.W., An extensive treatment of the agglomerate model for porous electrodes in molten carbonate fuel cells: II. Quantitative analysis of time dependent and steady-state model, *Electrochimica Acta* 42, 1997, 3585–3600.

13 Yuh, C.Y., Selman, J.R., The polarization of molten carbonate fuel cell electrodes: 1. Analysis of steady-state polarization data, *Journal of The Electrochemical Society* 138, 1991, 3642–3648.

14 Bednarz, M., Mechanistische Untersuchung und Modellierung der Kathodenreaktion in Karbonatbrennstoffzellen (MCFC), Dissertation, IPCH Universitdt Hamburg, 2002.

# 4
# Index Analysis of Models

*Kurt Chudej, Hans Josef Pesch, and Joachim Rang*

Index analysis of partial differential-algebraic equations (PDAE) – sometimes called singular PDEs – is a young research topic. It started with the introduction of (DAE-) indices based on MOL-discretised or Laplace- or Fourier-transformed versions of the PDAE and matured with the definition of algebraic, differential and perturbation indices of the PDAE itself (e.g. [6, 7, 9, 16, 18–20, 23, 24]).

Why index analysis?

An important prerequisite for the start of any numerical solution method on a partial differential equation (PDE) is the identification of correct and complete initial and boundary conditions. For (partial) differential-algebraic equations the identification of *consistent* initial and boundary conditions is especially important from a theoretical and numerical point of view.

Moreover, an index analysis can give hints for a better modelling approach; especially it can identify an unsuitable choice of variables.

Additionally, a (perturbation) index analysis gives an estimate of the accuracy of the computed numerical solution with respect to the underlying hierarchy of mathematical models. Moreover, the estimate of accuracy can be extended also to perturbed models, e.g. with respect to perturbed system constants.

Finally, the index influences the choice of a suitable numerical method.

We will compute the differential time index and the MOL index for some classes of 2D crossflow MCFC PDAE models, as well as a perturbation index for a simplified semilinear MCFC PDAE model.

## 4.1
## Differential Time Index

We consider the following class of PDAEs of mixed parabolic hyperbolic type: Find $\mathbf{u}(\tau, z)$ with time $\tau \in \bar{J}$, $J := (0, T)$ and spatial coordinate $z \in \bar{\Omega}$, $\Omega := (0, 1)^d$, $d \in \{1, 2, 3\}$, s.t.

$$A\mathbf{u}_\tau = \Psi(\mathbf{u}, \mathbf{u}_{z_1}, \dots, \mathbf{u}_{z_d}, \Delta\mathbf{u}, \tau, z) \tag{4.1}$$

*Molten Carbonate Fuel Cells.* Edited by Kai Sundmacher, Achim Kienle, Hans Josef Pesch, Joachim F. Berndt, and Gerhard Huppmann
Copyright © 2007 WILEY-VCH Verlag GmbH & Co. KGaA, Weinheim
ISBN: 978-3-527-31474-4

($A$ is a given constant matrix, e.g. $A = \operatorname{diag}(I, O)$) and initial conditions

$$A[\mathbf{u}(0, z) - \mathbf{g}(z)] = 0 \tag{4.2}$$

and suitable[1] nonlinear boundary conditions. We assume that the PDAE has a sufficiently smooth (classical) solution $\mathbf{u}$.

The following definition is a generalisation of [17], and [4], see [3]:

**Definition 1:** *If the matrix $A$ is regular, the (differential) time index is defined to be $v_\tau := 0$. Otherwise the (differential) time index $v_\tau$ is the smallest number of times; the PDAE must be differentiated with respect to time $\tau$ in order to determine $\mathbf{u}_\tau$ as a continuous function of $\tau$, $z$, $\mathbf{u}$ and spatial partial derivatives of $\mathbf{u}$.*

The knowledge of the time index is important for the choice of a suitable numerical solution method [9]. We consider the following 2D crossflow model of Heidebrecht, see Chapter 3 in this book, as well as certain simplifications. That is, $d = 2$, $\Gamma_{\text{west}} := \{(0, z_2) : z_2 \in [0, 1]\}$ and $\Gamma_{\text{south}} := \{(z_1, 0) : z_1 \in [0, 1]\}$.

Find $\mathbf{u} = (\theta_s, \theta_a, \theta_c, \chi_a, \chi_c, p_a, p_c, \gamma_a, \gamma_c, \Phi_a^L, \Phi_c^L, \Phi_c^S)$ subject to the PDAE:

*Heat equation in the solid:*

$$\frac{\partial \theta_s}{\partial \tau} = \lambda \Delta \theta_s + \varphi_1(\theta_s, \theta_a, \theta_c, \chi_a, \chi_c, p_a, p_c, \Phi_a^L, \Phi_c^L, \Phi_c^S), \quad \lambda \text{ positive constant} \tag{4.3}$$

*Advection equations in the gas streams:*

$$\frac{\partial \chi_{a,j}}{\partial \tau} = -\gamma_a \theta_a \frac{\partial \chi_{a,j}}{\partial z_1} + \varphi_{2,j}(\theta_s, \theta_a, \chi_a, p_a, \Phi_a^L), \quad j = 1, \ldots, 7 \tag{4.4a}$$

$$\frac{\partial \chi_{c,j}}{\partial \tau} = -\gamma_c \theta_c \frac{\partial \chi_{c,j}}{\partial z_2} + \varphi_{3,j}(\theta_s, \theta_c, \chi_c, p_c, \Phi_c^L, \Phi_c^S), \quad j = 1, \ldots, 7 \tag{4.4b}$$

$$\frac{\partial \theta_a}{\partial \tau} = -\gamma_a \theta_a \frac{\partial \theta_a}{\partial z_1} + \varphi_4(\theta_s, \theta_a, \chi_a, p_a, \Phi_a^L) \tag{4.4c}$$

$$\frac{\partial \theta_c}{\partial \tau} = -\gamma_c \theta_c \frac{\partial \theta_c}{\partial z_2} + \varphi_5(\theta_s, \theta_c, \chi_c, p_c, \Phi_c^L, \Phi_c^S) \tag{4.4d}$$

*Algebraic equations in the pores:*

$$0 = p_{a,j} - \varphi_{6,j}(p_a, \theta_s, \chi_a, \Phi_a^L), \quad j = 1, \ldots, 4 \tag{4.5a}$$

$$0 = p_{c,j} - \varphi_{7,j}(p_c, \theta_s, \chi_c, \Phi_c^L, \Phi_c^S), \quad j = 1, 2 \tag{4.5b}$$

---

1) Since part of the PDAE is of hyperbolic type only certain boundary conditions are allowed, see, e.g., [15].

*Degenerated PDEs for the molar flows:*

$$0 = \frac{\partial(\gamma_a \theta_a)}{\partial z_1} + \varphi_8(\theta_s, \theta_a, \chi_a, p_a, \Phi_a^L) \tag{4.6a}$$

$$0 = \frac{\partial(\gamma_c \theta_c)}{\partial z_2} + \varphi_9(\theta_s, \theta_c, \chi_c, p_c, \Phi_c^L, \Phi_c^S) \tag{4.6b}$$

*Integro PDAE system for the potentials:*

$$\frac{\partial \Phi_a^L}{\partial \tau} = [i_a(p_a, \theta_s, \Phi_a^L) - i]/c_a, \quad c_a, c_c \text{ constants} \tag{4.7a}$$

$$\frac{\partial \Phi_c^L}{\partial \tau} = [i_a(p_a, \theta_s, \Phi_a^L) - i]/c_a + [i_e(\Phi_a^L, \Phi_c^L) - i]/c_c \tag{4.7b}$$

$$\frac{d\Phi_c^S}{d\tau} = \varphi_{10}\left( \int_\Omega i_a(p_a, \theta_s \Phi_a^L)\, dz, \int_\Omega i_c(p_c, \theta_s, \Phi_c^L, \Phi_c^S)\, dz, \int_\Omega i_e(\Phi_a^L, \Phi_c^L)\, dz, I_{\text{cell}} \right) \tag{4.7c}$$

Additionally, initial conditions at $\tau = 0$ of type (4.2) for $(\theta_s, \theta_a, \theta_c, \chi_a, \chi_c, \Phi_a^L, \Phi_c^L, \Phi_c^S)$, and Neumann boundary conditions for the solid temperature

$$\theta_s|_{\partial\Omega} = 0 \tag{4.8}$$

are given. (The consistent initial conditions for $\gamma_a$, $\gamma_c$, $p_a$, $p_c$ at $\tau = 0$ are already determined through (4.2) and (4.5a), (4.5b), (4.6a) and (4.6b).

Linear boundary conditions at $\Gamma_{\text{west}}$ are prescribed:

$$\theta_a|_{\Gamma_{\text{west}}} = \theta_{a,\text{in}}(\tau), \quad \chi_a|_{\Gamma_{\text{west}}} = \chi_{a,\text{in}}(\tau), \quad \gamma_a|_{\Gamma_{\text{west}}} = \gamma_{a,\text{in}}(\tau) \tag{4.9}$$

Nonlinear boundary conditions at $\Gamma_{\text{south}}$ are given by the following DAE:

$$\theta_c|_{\Gamma_{\text{south}}} = \theta_m(\tau), \quad \chi_c|_{\Gamma_{\text{south}}} = \chi_m(\tau), \quad \gamma_c|_{\Gamma_{\text{south}}} = \gamma_m(\tau) \tag{4.10}$$

where

$$\frac{d\chi_{m,j}}{d\tau} = \varphi_{12,j}\left( \chi_m, \theta_m, \int_0^1 \chi_a|_{z_1=1}\, dz_2, \int_0^1 \theta_a|_{z_1=1}\, dz_2, \int_0^1 \gamma_a|_{z_1=1}\, dz_2, \chi_{\text{air}}, \lambda_{\text{air}} \right) \tag{4.11a}$$

$$\frac{d\theta_m}{d\tau} = \varphi_{13}\left( \theta_m, \int_0^1 \chi_a|_{z_1=1}\, dz_2, \int_0^1 \theta_a|_{z_1=1}\, dz_2, \int_0^1 \gamma_a|_{z_1=1}\, dz_2, \theta_{\text{air}}, \chi_{\text{air}}, \lambda_{\text{air}} \right) \tag{4.11b}$$

$$\gamma_m = \varphi_{14}\left( \theta_m, \int_0^1 \chi_a|_{z_1=1}\, dz_2, \int_0^1 \theta_a|_{z_1=1}\, dz_2, \int_0^1 \gamma_a|_{z_1=1}\, dz_2, \theta_{\text{air}}, \chi_{\text{air}}, \lambda_{\text{air}} \right) \tag{4.11c}$$

with the initial conditions $\chi_{m,j}(0) = \chi_{m,j,0}$, $\theta_m(0) = \theta_{m,0}$.

The given functions $i = \varphi_{11}(i_a, i_c, i_e, \int_\Omega i_a\,dz, \int_\Omega i_c\,dz, \int_\Omega i_e\,dz, I_{cell})$ and $\varphi_j$, $i_a$, $i_e$, $i_c$, $I_{cell}$ as well as $\theta_{a,in}$, $\chi_{a,in}$, $\gamma_{a,in}$, $\theta_{air}$, $\chi_{air}$, $\lambda_{air}$ are assumed to be sufficiently smooth.

In the following, we assume that by the coordinate transformation $v = \gamma_a\theta_a$ and $\bar{v} = \gamma_c\theta_c$ the variables $\gamma_a$ and $\gamma_c$ are replaced by $v$ and $\bar{v}$ (the same is done for the boundary conditions).

We first consider a simplified 2D model of the MCFC. We neglect the partial pressures in the pores and assume the electrical potentials to be constant. Then Eqs. (4.3), (4.4a)–(4.4d), (4.6a), (4.6b) together with simplified linear boundary conditions can be written as:

Find functions $u(\tau, z)$ (temperature of the solid), $v(\tau, z) > 0$, $\bar{v}(\tau, z) > 0$ (temperature times molar flow of the gas in the anode and cathode gas channel) and $w(\tau, z)$, $\bar{w}(\tau, z)$ (temperature and molar fractions of the gas in the anode and cathode gas channel) s.t. the PDAE (with positive constant $\lambda$) on $J \times \Omega$

$$u_\tau = \lambda\Delta u + \psi_1(u, w, v, \bar{w}, \bar{v}) \tag{4.12}$$

$$w_\tau = -vw_{z_1} + \psi_2(u, w, v) \tag{4.13}$$

$$0 = v_{z_1} + \psi_3(u, w) \tag{4.14}$$

$$\bar{w}_\tau = -\bar{v}\bar{w}_{z_2} + \psi_4(u, \bar{w}, \bar{v}) \tag{4.15}$$

$$0 = \bar{v}_{z_2} + \psi_5(u, \bar{w}) \tag{4.16}$$

with initial conditions

$$u(0, z) = g_1(z), \quad w(0, z) = g_2(z), \quad \bar{w}(0, z) = g_4(z) \tag{4.17}$$

and boundary conditions

$$\left.\frac{\partial u}{\partial n}\right|_{\partial\Omega} = 0, \tag{4.18}$$

$$w(\tau, 0, z_2) = w_{west}(\tau, z_2), \quad v(\tau, 0, z_2) = v_{west}(\tau, z_2), \tag{4.19}$$

$$\bar{w}(\tau, z_1, 0) = \bar{w}_{south}(\tau, z_1), \quad \bar{v}(\tau, z_1, 0) = \bar{v}_{south}(\tau, z_1) \tag{4.20}$$

We assume that all the given functions $\psi_i$, $g_i$, $w_{west}$, $v_{west}$, $\bar{w}_{south}$, $\bar{v}_{south}$ are continuous with respect to all arguments. We assume that the given functions $\psi_3$ and $\psi_5$ are continuously differentiable with respect to all arguments. We assume that the given functions $v_{west}$ and $\bar{v}_{south}$ are continuously differentiable with respect to $\tau$. We assume that $g_2(z_1 = 0, z_2) = w_{west}(\tau = 0, z_2)$ and $g_4(z_1, z_2 = 0) = \bar{w}_{south}(\tau = 0, z_1)$.

**Theorem 2:** The differential time index of the PDAE (4.12)–(4.20) is $v_\tau = 1$.

*Proof:* For a shorter notation scalar $w$, $\bar{w}$ are assumed. Partial differentiation with respect to time of Eqs. (4.14) and (4.16) yields

$$0 = v_{\tau z_1} + \frac{\partial \psi_3}{\partial u} u_\tau + \frac{\partial \psi_3}{\partial w} w_\tau, \quad 0 = \bar{v}_{\tau z_2} + \frac{\partial \psi_5}{\partial u} u_\tau + \frac{\partial \psi_5}{\partial \bar{w}} \bar{w}_\tau$$

Plugging on the right-hand side of Eqs. (4.12), (4.13), (4.15) yields

$$\frac{\partial}{\partial z_1} v_\tau = \alpha(u, w, v, \bar{w}, \bar{v}, \Delta u, w_{z_1}), \quad \frac{\partial}{\partial z_2} \bar{v}_\tau = \bar{\alpha}(u, w, v, \bar{w}, \bar{v}, \Delta u, \bar{w}_{z_2})$$

Due to the assumptions $\alpha$ and $\bar{\alpha}$ are continuous with respect to all arguments. Therefore the formulae

$$v_\tau(\tau, z) = \frac{\partial}{\partial \tau} v_{\text{west}}(\tau, z_2) + \int_0^{z_1} \alpha(u, w, v, \bar{w}, \bar{v}, \Delta u, w_{z_1})|_{(\tau, \tilde{z}_1, z_2)} \, d\tilde{z}_1, \tag{4.21}$$

$$\bar{v}_\tau(\tau, z) = \frac{\partial}{\partial \tau} \bar{v}_{\text{south}}(\tau, z_1) + \int_0^{z_2} \bar{\alpha}(u, w, v, \bar{w}, \bar{v}, \Delta u, \bar{w}_{z_2})|_{(\tau, z_1, \tilde{z}_2)} \, d\tilde{z}_2 \tag{4.22}$$

yield the time index $v_\tau = 1$, which also holds for the vectors $w$, $\bar{w}$.

**Theorem 3:** The differential time index of the PDAE (4.12)–(4.16) is $v_\tau = 1$ if linear boundary conditions (4.18)–(4.19), nonlinear boundary conditions

$$\bar{w}(\tau, z_1, 0) = \phi_1\left(\xi(\tau), \int_0^1 w(\tau, 1, z_2) \, dz_2, \int_0^1 v(\tau, 1, z_2) \, dz_2\right), \tag{4.23a}$$

$$\bar{v}(\tau, z_1, 0) = \phi_2\left(\xi(\tau), \int_0^1 w(\tau, 1, z_2) \, dz_2, \int_0^1 v(\tau, 1, z_2) \, dz_2\right) \tag{4.23b}$$

with a given continuously differentiable function $\xi(\tau)$ and initial conditions (4.17) are considered. Here, the given function $\phi_1$ is assumed to be continuous, and the given function $\phi_2$ to be continuously differentiable with respect to all arguments.

*Proof:* The only change of the proof of Theorem 2 is that one has to replace Eq. (4.22) by

$$\bar{v}_\tau(\tau, z) = \left(\frac{d\xi(\tau)}{dt}, \int_0^1 -vw_{z_1} + \psi_2(u, w, v)|_{(\tau, 1, z_2)} \, dz_2, \int_0^1 v_\tau(\tau, 1, z_2) \, dz_2\right)$$

$$\cdot \, \text{grad} \, \phi_2\left(\xi(\tau), \int_0^1 w(\tau, 1, z_2) \, dz_2, \int_0^1 v(\tau, 1, z_2) \, dz_2\right)$$

$$+ \int_0^{z_2} \bar{\alpha}(u, w, v, \bar{w}, \bar{v}, \Delta u, \bar{w}_{z_2})|_{(\tau, z_1, \tilde{z}_2)} \, d\tilde{z}_2 \tag{4.24}$$

while inserting Eq. (4.21) for $v_\tau$.

Finally, we return to the full 2D model of the MCFC:

**Theorem 4:** The differential time index of the PDAE (4.3)–(4.7c) is $v_\tau = 1$ if linear boundary conditions of type (4.18)–(4.20) with $w = (\theta_a, \chi_a)$, $\bar{w} = (\theta_c, \chi_c)$, $v = (p_a, \gamma_a)$, $\bar{v} = (p_c, \gamma_c)$ and initial conditions (4.2) are considered.

*Proof:* Investigation of the functions $\varphi_{6,j}$, $\varphi_{7,j}$ yields that Eqs. (4.5a) and (4.5b) for the partial pressures in the pores are of time index 1, since $\frac{\partial \varphi_6}{\partial p_a} - I$ and $\frac{\partial \varphi_7}{\partial p_c} - I$ are regular. Equations (4.7a)–(4.7c) for the potentials are obviously of time index zero, if the electrical density $i$ is considered only as an auxiliary function. Application of the ideas in the proof of Theorem 2 finishes the proof.

Investigation of the fuel cell model of Section 2 including the nonlinear boundary conditions is more complicated:

**Theorem 5:** We consider the full model, i.e. the PDAE (4.3)–(4.11c) with initial conditions (4.2). If we assume that the functions $\int_0^1 \chi_{a,j}(\tau, 1, z_2)\, dz_2$, $\int_0^1 \theta_a(\tau, 1, z_2)\, dz_2$, $\int_0^1 \gamma_a(\tau, 1, z_2)\, dz_2$ are continuously differentiable with respect to time and that $\theta_{\mathrm{air}}(\tau)$, $\chi_{\mathrm{air}}(\tau)$, $\lambda_{\mathrm{air}}(\tau)$ are continuously differentiable, then the differential time index of the full model is $v_\tau = 1$.

*Proof:* The solution of Eqs. (4.10)–(4.11c) can be written as a nonlinear boundary condition (4.23a), (4.23b). Under the above assumptions the functions $\xi(\tau) = (\theta_{\mathrm{air}}(\tau), \chi_{\mathrm{air}}(\tau), \lambda_{\mathrm{air}}(\tau))$ and $\chi_m(\tau)$, $\theta_m(\tau)$ are continuously differentiable; therefore, $\phi_1$ and $\phi_2$ and all of their arguments are continuously differentiable. Application of Theorems 3 and 4 finishes the proof.

**Theorem 6:** Theorems 2 and 5 remain true if instead of the variables $v$ and $\bar{v}$ the variables $\gamma_a$ and $\gamma_c$ with their Eqs. (4.6a), (4.6b) are used.

## 4.2
## MOL Index

After an equidistant semi-discretization in space of one of the several previous nonlinear MCFC PDAE models we get a semi-explicit differential-algebraic initial value problem in time $\tau$ for a huge dimensional variable $\mathbf{U}(\tau)$:

$$\mathscr{A} \frac{d}{d\tau} \mathbf{U}(\tau) + b(\mathbf{U}(\tau), \Xi(\tau)) = 0, \quad \mathscr{A}(\mathbf{U}(0) - \mathbf{U}_0) = 0 \tag{4.25}$$

The Laplace operator is approximated by the five-point star. The first spatial partial derivatives are approximated by appropriate upwind formulae. The diagonal matrix $\mathscr{A}$ has zeros or ones in the diagonal.

$\mathbf{U}$ includes the component functions along the lines associated with the space discretization as well as the functions $\chi_m(\tau)$ and $\theta_m(\tau)$. $\Xi(\tau)$ denotes the boundary functions $w_{\mathrm{west}}(\tau) = (\theta_{a,\mathrm{in}}(\tau), \chi_{a,\mathrm{in}}(\tau))$, $v_{\mathrm{west}}(\tau) = \gamma_{a,\mathrm{in}}(\tau)$ and $\chi_{\mathrm{air}}(\tau)$, $\theta_{\mathrm{air}}(\tau)$, $\lambda_{\mathrm{air}}(\tau)$, which are either given or can be used for control purposes.

**Theorem 7:** The MOL index of the semi-explicit DAEs derived from all previous MCFC PDAE models is $\nu_{\text{MOL}} = 1$.

The proof is based on the upwind formulae with the known wind direction and the correctly posed boundary conditions of the PDAE; compare Chudej et al. [4], Chap. 4, for the 1D case.

The MOL index $\nu_{\text{MOL}} = 1$ coincides with the perturbation index of the semi-explicit DAE. So the optimal control problem of finding time-dependent optimal boundary controls s.t. the DAE constraint is well posed.

## 4.3
## Perturbation Index

From a numerical point of view a perturbation index is more important, since it characterises the well-posedness of the PDAE boundary value problem, see e.g. [7], [16], [11].

Later we want to control the PDAE system by choosing some of the boundary functions $\Xi(\tau)$ as (optimal) controls; therefore; knowledge about the distance between solutions of perturbed problems is especially valuable.

As a first result we present the perturbation index for a semilinear version of the PDAE (4.12)–(4.20) with $\mathbf{u} = (u, w, v, \bar{w}, \bar{v}) = (u_1, \ldots, u_5)$:

$$u_\tau = \lambda \Delta u + \hat{\psi}_1(u, w, v, \bar{w}, \bar{v}), \quad \lambda \text{ positive constant,} \tag{4.26}$$

$$w_\tau = -b(\tau, z)w_{z_1} + \hat{\psi}_2(u, w, v), \quad b(\tau, z) > 0, \tag{4.27}$$

$$0 = v_{z_1} + \hat{\psi}_3(u, w), \tag{4.28}$$

$$\bar{w}_\tau = -\bar{b}(\tau, z)\bar{w}_{z_2} + \hat{\psi}_4(u, \bar{w}, \bar{v}), \quad \bar{b}(\tau, z) > 0, \tag{4.29}$$

$$0 = \bar{v}_{z_2} + \hat{\psi}_5(u, \bar{w}), \tag{4.30}$$

with $\hat{\psi}_j(\mathbf{u}) = -\sum_i c_{ji}(\tau, z)u_i + f_j(\tau, z)$ with initial conditions (4.17) and linear boundary conditions (4.18)–(4.20).

## 4.3.1
## Transformation to Homogenous Dirichlet Boundary Conditions

For a shorter notation scalar $w$, $\bar{w}$ are assumed. We apply a coordinate transformation $w^{\text{new}}(\tau, z) := w(\tau, z) - w_{\text{west}}(\tau)$ and $v^{\text{new}}(\tau, z) := v(\tau, z) - v_{\text{west}}(\tau)$ and $\bar{w}^{\text{new}}(\tau, z) := \bar{w}(\tau, z) - \bar{w}_{\text{south}}(\tau)$ and $\bar{v}^{\text{new}}(\tau, z) := \bar{v}(\tau, z) - \bar{v}_{\text{south}}(\tau)$ to the semilinear PDAE (4.26)–(4.30), (4.17)–(4.20). The new solution functions $(u, w^{\text{new}}, v^{\text{new}}, \bar{w}^{\text{new}}, \bar{v}^{\text{new}})$ satisfy again Eqs. (4.26)–(4.30), but with a new right-hand side $f^{\text{new}}$ defined by $f_j^{\text{new}} = f_j - c_{j2}w_{\text{west}} - c_{j3}v_{\text{west}} - c_{j4}\bar{w}_{\text{south}} - c_{j5}\bar{v}_{\text{south}}$, initial conditions (4.17) and *homogeneous* linear boundary conditions (4.18)–(4.20), i.e. $w_{\text{west}}^{\text{new}} = 0$, $v_{\text{west}}^{\text{new}} = 0$, $\bar{w}_{\text{south}}^{\text{new}} = 0$, $\bar{v}_{\text{south}}^{\text{new}} = 0$. In the following we neglect the superscript 'new'.

We assume that the coefficients $b, \bar{b}, c_{ij} : J \times \Omega \to \mathbb{R}$ have the properties $b, \bar{b}, c_{ij} \in C(J, L_\infty(\Omega))$, and $f_i \in C(J, L^2(\Omega))$.

We assume that a sufficiently smooth (classical) solution $\mathbf{u}$ exists.

### 4.3.2
### Abstract Problem

First we multiply Eqs. (4.26)–(4.30) with test functions $\tilde{u} \in H^1(\Omega)$, $\tilde{w}, \tilde{v} \in H^1_{\text{west}}(\Omega) := \{v \in H^1(\Omega) : v|_{\Gamma_{\text{west}}} = 0\}$, $\tilde{\bar{w}}, \tilde{\bar{v}} \in H^1_{\text{south}}(\Omega) := \{v \in H^1(\Omega) : v|_{\Gamma_{\text{south}}} = 0\}$. Moreover, we set $V := H^1(\Omega) \times (H^1_{\text{west}}(\Omega))^2 \times (H^1_{\text{south}}(\Omega))^2$, $\|u(\tau, z)\|^2 = \int_\Omega |u(\tau, z)|^2 \, dz$ and $\|\mathbf{u}\|^2 = \sum_i \|u_i\|^2$.

Integrating over the domain $\Omega$ yields with integration by parts the weak formulation:

$$\langle u_\tau, \tilde{u} \rangle + \lambda \langle \nabla u, \nabla \tilde{u} \rangle + \langle c_{11} u + c_{12} w + c_{13} v + c_{14} \bar{w} + c_{15} \bar{v}, \tilde{u} \rangle = \langle f_1, \tilde{u} \rangle$$

$$\langle w_\tau, \tilde{w} \rangle + \langle b w_{z_1}, \tilde{w} \rangle + \langle c_{21} u + c_{22} w + c_{23} v, \tilde{w} \rangle = \langle f_2, \tilde{w} \rangle$$

$$\langle v_{z_1}, \tilde{v} \rangle + \langle c_{31} u + c_{32} w, \tilde{v} \rangle = \langle f_3, \tilde{v} \rangle$$

$$\langle \bar{w}_\tau, \tilde{\bar{w}} \rangle + \langle \bar{b} \bar{w}_{z_2}, \tilde{\bar{w}} \rangle + \langle c_{41} u + c_{44} \bar{w} + c_{45} \bar{v}, \tilde{\bar{w}} \rangle = \langle f_4, \tilde{\bar{w}} \rangle$$

$$\langle \bar{v}_{z_2}, \tilde{\bar{v}} \rangle + \langle c_{51} u + c_{54} \bar{w}, \tilde{\bar{v}} \rangle = \langle f_5, \tilde{\bar{v}} \rangle$$

With $\mathbf{u} = (u, w, v, \bar{w}, \bar{v}) = (u_1, u_2, u_3, u_4, u_5)$ and the operators $\mathscr{A}, \mathscr{B} : V \to V^*$ and $\mathbf{f} \in V^*$

$$\langle \mathscr{A} \mathbf{u}, \tilde{\mathbf{u}} \rangle := \langle u, \tilde{u} \rangle + \langle w, \tilde{w} \rangle + \langle \bar{w}, \tilde{\bar{w}} \rangle, \quad \langle \mathbf{f}, \tilde{\mathbf{u}} \rangle := \sum_i \langle f_i, \tilde{u}_i \rangle,$$

$$\langle \mathscr{B} \mathbf{u}, \tilde{\mathbf{u}} \rangle := \lambda \langle \nabla u, \nabla \tilde{u} \rangle + \langle b w_{z_1}, \tilde{w} \rangle + \langle \bar{b} \bar{w}_{z_2}, \tilde{\bar{w}} \rangle + \sum_{i,j} \langle c_{ij} u_j, \tilde{u}_i \rangle,$$

we get the abstract DAE (ADAE) in $V^*$ for the unknown element $\mathbf{u} \in V$ with initial conditions:

$$\mathscr{A} \mathbf{u}_\tau + \mathscr{B} \mathbf{u} = \mathbf{f}, \quad \mathscr{A}(\mathbf{u} - \mathbf{u}_0) = 0 \qquad (4.31)$$

### 4.3.3
### Perturbation Index

Let us consider the ADAE (4.31) $\mathscr{A} \mathbf{u}_\tau + \mathscr{B} \mathbf{u} = \mathbf{f}$ with solution $\mathbf{u}$ and the perturbed ADAE $\mathscr{A} \hat{\mathbf{u}}_\tau + \mathscr{B} \hat{\mathbf{u}} = \mathbf{f} + \delta$ with solution $\hat{\mathbf{u}}$. Then $\varepsilon := \hat{\mathbf{u}} - \mathbf{u}$ solves $\mathscr{A} \varepsilon_\tau + \mathscr{B} \varepsilon = \delta$. Due to the transformation to homogeneous boundary conditions, a perturbation in the Dirichlet boundary conditions can be interpreted as an additional perturbation in $\delta$.

**Definition 8:** *If the inequality*

$$\|\hat{\mathbf{u}}(\tau) - \mathbf{u}(\tau)\| \le c(\|\hat{\mathbf{u}}(0) - \mathbf{u}(0)\| + \sup_{t \in J} \|\boldsymbol{\delta}(t)\|_{V^*})$$

*holds, then the ADAE has perturbation index* $i_p = 1$.

A more detailed definition can be found in [22–24].

**Theorem 9:** The perturbation index of the semilinear 2D MCFC PDAE model (4.26)–(4.30), (4.17)–(4.20) is $v_p = 1$.

This is a good result from an engineering point of view. Perturbations in the boundary functions – which may act as boundary controls – influence the solution only by themself (no derivatives of the perturbations appear!).

In the following we give a proof of Theorem 9.

### 4.3.4
### Garding-Type Inequality

Due to the hyperbolic equations the theory about the Garding-type inequality from Rang and Angermann [23] cannot be adopted, but a weaker result can still be proved, see [22, 24]:

There exists a constant $\tilde{c} > 0$ such that for all $\mathbf{u} \in V$ the following holds:

$$\langle \mathscr{B}\mathbf{u}, \mathbf{u} \rangle \ge \lambda \|\nabla u\|^2 - \tilde{c}\|\mathbf{u}\|^2 \tag{4.32}$$

This is due to

$$2\langle bw_{z_1}, w \rangle = \left\langle b, \frac{\partial}{\partial z_1} w^2 \right\rangle \ge \inf_{z \in \Omega} b \int_0^1 \int_0^1 \frac{\partial}{\partial z_1} w^2 \, dz_1 \, dz_2$$

$$\stackrel{(4.19)}{=} \inf_{z \in \Omega} b(\tau, z) \int_0^1 [w(\tau, 1, z_2)]^2 \, dz_2 \ge 0$$

and $\langle \bar{b}\bar{w}_{z_2}, \bar{w} \rangle \ge 0$ and $\langle c_{ij}u_j, u_i \rangle \ge -\gamma_{ij}\|u_j\| \|u_i\|$ with $\tilde{c} = \sup_{\tau \in J} \|(\gamma_{ij})_{ij}\|_2$ and $\gamma_{ij} = \sup_{z \in \Omega} |c_{ij}|$.

### 4.3.5
### Estimate for $v$ and $\bar{v}$

$$|v(\tau, z)|^2 \stackrel{(4.19)}{\le} \left[ \int_0^{z_1} |v_{z_1}(\tau, \tilde{z}_1, z_2)| \, d\tilde{z}_1 \right]^2 \le \left[ \int_0^1 |v_{z_1}(\tau, z_1, z_2)| \, dz_1 \right]^2$$

$$\le \int_0^1 |v_{z_1}(\tau, z_1, z_2)|^2 \, dz_1$$

$$\int_0^1 |v(\tau, z)|^2 \, dz_2 \leq \int_\Omega |v_{z_1}(\tau, z_1, z_2)|^2 \, dz = \|v_{z_1}\|^2$$

$$\int_\Omega |v(\tau, z)|^2 \, dz = \|v\|^2 \leq \int_0^1 \|v_{z_1}\|^2 \, dz_1 \leq \|v_{z_1}\|^2$$

Therefore with $\gamma_3 = \sup_{\tau \in J}\{\gamma_{31}, \gamma_{32}\}$ and $\gamma_5 = \sup_{\tau \in J}\{\gamma_{51}, \gamma_{54}\}$ the following holds:

$$\|v\| \leq \|v_{z_1}\| \overset{(4.28)}{\leq} \|f_3\| + \gamma_3(\|u\| + \|w\|) \tag{4.33}$$

$$\|\bar{v}\| \leq \|\bar{v}_{z_2}\| \overset{(4.30)}{\leq} \|f_5\| + \gamma_5(\|u\| + \|\bar{w}\|) \tag{4.34}$$

### 4.3.6
### Estimate for *u*, *w* and *w̄* with Garding-Type Inequality

$$\langle \mathbf{f}, \mathbf{u} \rangle - \langle \mathscr{A}\mathbf{u}_\tau, \mathbf{u} \rangle = \langle \mathscr{B}\mathbf{u}, \mathbf{u} \rangle \geq \lambda\|\nabla u\|^2 - \tilde{c}\|u\|^2$$

$$\Rightarrow \langle \mathscr{A}\mathbf{u}_\tau, \mathbf{u} \rangle + \lambda\|\nabla u\|^2 - \tilde{c}\|u\|^2 \leq \langle \mathbf{f}, \mathbf{u} \rangle \leq \|u\|^2 + \frac{1}{4}\|\mathbf{f}\|^2$$

Therefore after neglecting $\lambda\|\nabla u\|^2$ we get

$$\langle u_\tau, u \rangle + \langle w_\tau, w \rangle + \langle \bar{w}_\tau, \bar{w} \rangle - (\tilde{c}+1)\|u\|^2 \leq \frac{1}{4}\|\mathbf{f}\|^2$$

Due to $2\langle u_\tau, u \rangle = \frac{d}{d\tau}\|u\|^2$ we get

$$\frac{1}{2}\frac{d}{d\tau}(\|u\|^2 + \|w\|^2 + \|\bar{w}\|^2) - (\tilde{c}+1)(\|u\|^2 + \|w\|^2 + \|\bar{w}\|^2)$$

$$\leq \frac{1}{4}\|\mathbf{f}\|^2 + (\tilde{c}+1)(\|v\|^2 + \|\bar{v}\|^2)$$

$$\overset{(4.33, 4.34)}{\leq} \frac{1}{4}\|\mathbf{f}\|^2 + (\tilde{c}+1)[\|f_3\| + \gamma_3(\|u\| + \|w\|)]^2$$

$$+ (\tilde{c}+1)[\|f_5\| + \gamma_5(\|u\| + \|\bar{w}\|)]^2$$

$$\leq \frac{C}{2}\|\mathbf{f}\|^2 + \frac{D}{2}(\|u\|^2 + \|w\|^2 + \|\bar{w}\|^2)$$

Therefore with $\alpha = 2(\tilde{c}+1) + D$

$$\frac{d}{d\tau}(\|u\|^2 + \|w\|^2 + \|\bar{w}\|^2) - \alpha(\|u\|^2 + \|w\|^2 + \|\bar{w}\|^2) \leq C\|\mathbf{f}\|^2$$

Multiplication by $e^{-\alpha\tau}$ and integration over $[0, \tau] \subseteq \bar{J}$ and multiplication with $e^{\alpha\tau}$ yields

$$\left(\|u(\tau, z)\|^2 + \|w(\tau, z)\|^2 + \|\bar{w}(\tau, z)\|^2\right)$$

$$\leq e^{\alpha\tau}\left(\|g_1(z)\|^2 + \|g_2(z)\|^2 + \|g_4(z)\|^2\right) + \int_0^\tau C e^{\alpha(\tau-s)} \|\mathbf{f}(s, z)\|^2 \, ds \quad (4.35)$$

A slight rearrangement of Eqs. (4.33), (4.34), (4.35) completes the proof of Theorem 9.

## 4.4
## Conclusion

All computed indices of the various models have index 1. Therefore, we know that the computed numerical solution of the MOL–DAE is accurate (Theorem 7) and we can also expect[2] that it is accurate with respect to the PDAE model. Moreover, the same can be expected with respect to perturbed PDAE models, possibly including small neglected physical or chemical effects as well as perturbations in system parameters.

### Bibliography

1 Angermann, L., Rang, J., Perturbation index of linear partial differential-algebraic equations with a hyperbolic part. Mathematik-Bericht Nr. 2005/7, Institut für Mathematik, Technische Universität Clausthal, 2005.

2 Büskens, C., Optimierungsmethoden und Sensitivitätsanalyse für optimale Steuerprozesse mit Steuer- und Zustands-Beschränkungen. Dissertation, Universität Münster, 1998.

3 Chudej, K., Index Analysis for Singular PDE Models of Fuel Cells. In: A. Di Bucchianico et al. (Eds.). *Progress in Industrial Mathematics at ECMI 2004.* Springer, Berlin, 2006, pp. 212–216.

4 Chudej, K., Heidebrecht, P., Petzet, V., Scherdel, S., Schittkowski, K., Pesch, H.J., Sundmacher, K., Index analysis and numerical solution of a large scale nonlinear PDAE system describing the dynamical behaviour of molten carbonate fuel cells, *Z. Angew. Math. Mech.* 85(2), 2005, 132–140.

5 Chudej, K., Sternberg, K., Pesch, H.J., Simulation and optimal control of molten carbonate fuel cells. In: I. Troch, F. Breitenecker (Eds.). *Proc. of 5th MATHMOD Vienna,* ARGESIM Report No. 30, ARGESIM-Verlag, Wien, 2006.

6 Campbell, S.L., Marszalek, W., ODE/DAE integrators and MOL problems,

2) The differential time index of the MCFC PDAE is 1, and we expect an analogous behaviour for PDAEs of type (4.1) as for semi-explicit DAEs, where it is known that for sufficiently large perturbation families the perturbation index is greater than or equal to the differential index [8, 10].

*Z. Angew. Math. Mech.* 76(Suppl. 1), 1996, S251–S254.

**7** Campbell, S.L., Marszalek, W., The index of an infinite dimensional implicit system, *Math. Comput. Model. Dyn. Syst.* 5(1), 1999, 18–42.

**8** Deuflhard, P., Bornemann, F., Numerische Mathematik II. 2. Aufl., W. de Gruyter, Berlin, 2002.

**9** Günther, M., Wagner, Y., Index concepts for linear mixed systems of differential-algebraic and hyperbolic-type equations, *SIAM J. Sci. Comput.* 22, 2000, 1610–1629.

**10** Hairer, E., Wanner, G., *Solving Ordinary Differential Equations II, Stiff and Differential-Algebraic Problems*, 2nd revised edn., Springer, Berlin, 1996.

**11** Hanke, M., Olsson, K., Strömgren, M., Stability analysis of a degenerate hyperbolic system modelling a heat exchanger. Report TRITA-NA-0518, Royal Institute of Technology (KTH) Stockholm, 2005.

**12** Heidebrecht, P., *Modelling, Analysis and Optimisation of a Molten Carbonate Fuel Cell with Direct Internal Reforming (DIR-MCFC)*. VDI Fortschritt Berichte, Reihe 3, Nr. 826, VDI Verlag, Düsseldorf, 2005.

**13** Heidebrecht, P., Sundmacher, K., Dynamic modeling and simulation of a countercurrent molten carbonate fuel cell (MCFC) with internal reforming, *Fuel Cells* 3–4, 2002, 166–180.

**14** Heidebrecht, P., Sundmacher, K., Molten carbonate fuel cell (MCFC) with internal reforming: model-based analysis of cell dynamics, *Chemical Engineering Science* 58, 2003, 1029–1036.

**15** Kröner, D., *Numerical Schemes for Conservation Laws*. Wiley, Teubner, Chichester, 1997.

**16** Lamour, R., März, R., Tischendorf, C., PDAEs and further mixed systems as abstract differential algebraic systems. Preprint 01-11, Institut für Mathematik, Humboldt-Universität zu Berlin, 2001.

**17** Lucht, W., Debrabant, K., On quasi-linear PDAEs with convection: applications, indices, numerical solution, *Appl. Numer. Math.* 42, 2002, 297–314.

**18** Lucht, W., Strehmel, K., Discretization based indices for semilinear partial differential algebraic equations, *Appl. Numer. Math.* 28, 1998, 371–386.

**19** Lucht, W., Strehmel, K., Eichler-Liebenow, C., Indexes and special discretization methods for linear partial differential algebraic equations, *BIT Numerical Mathematics* 39(3), 1999, 484–512.

**20** Martinson, W.S., Barton, P.I., A differentiation index for partial differential equations, *SIAM J. Sci. Comput.* 21(6), 2000, 2295–2315.

**21** Pesch, H.J., Sternberg, K., Chudej, K., Towards the numerical solution of a large scale PDAE constrained optimization problem arising in molten carbonate fuel cell modeling. In: G. Di Pillo, M. Roma (Eds.). *Large Scale Nonlinear Optimization*. Berlin, Springer, 2006, 243–253.

**22** Rang, J., Stability estimates and numerical methods for degenerate parabolic differential equations. Papierflieger, Clausthal-Zellerfeld, 2005.

**23** Rang, J., Angermann, L., Perturbation index of linear partial differential-algebraic equations, *Appl. Numer. Math.* 53(2–4), 2005, 437–456.

**24** Rang, J., Chudej, K., A perturbation index for a singular PDE model of a fuel cell. Mathematik-Bericht Nr. 2004/6, Institut für Mathematik, Technische Universität Clausthal, 2004.

**25** Sternberg, K., Simulation, Optimale Steuerung und Sensitivitäts-analyse einer Schmelzkarbonat-Brennstoffzelle. Dissertation, Universität Bayreuth, 2007.

# 5
# Parameter Identification

*Matthias Gundermann and Kai Sundmacher*

To solve the MCFC model from Chapter 3 specific values have to be assigned to the model parameters. The aim of this chapter is to determine these parameter values so that the model reflects the real behaviour of the MCFC as good as possible.

Therefore several measurements on a MCFC power plant were conducted, which are explained in Section 5.1. The subsequent parameter estimation is a very complex optimisation problem with a large number of parameters and a set of highly non-linear and coupled PDEs as equality constraints resulting from the MCFC model equations. To carry out this optimisation a specific strategy has been developed which is presented in Section 5.2.

Finally, the results of the parameter identification are summarised and analysed in Section 5.3.

## 5.1
## Experimental Work

For the experimental work the fuel cell power plant HotModule was used that is operated by the project partner IPF at the University Hospital in Magdeburg (see Chapter 2).

As this HotModule was installed as a test plant it is equipped with much more measuring instruments than such a plant would usually have in a later serial production. Nevertheless measurements are still limited compared to a laboratory system. That means the kind of measuring instruments as well as the type and the position of the sensors are determined by the manufacturer and they cannot be changed. It is obvious that the installation of the sensors follows the requirements of process control and safety so that the chosen configuration is not necessarily advantageous for the estimation of the model parameters. Thus some important measurements are missing and the accuracy is not always sufficient.

An important aspect of the experimental work is the interpretation of the measurements with regard to the parameter identification. As explained in Chapter 3

*Molten Carbonate Fuel Cells.* Edited by Kai Sundmacher, Achim Kienle, Hans Josef Pesch, Joachim F. Berndt, and Gerhard Huppmann
Copyright © 2007 WILEY-VCH Verlag GmbH & Co. KGaA, Weinheim
ISBN: 978-3-527-31474-4

the MCFC model describes a single cell whereas the real HotModule contains a stack of 342 cells. Because there are distinct differences between the individual cells, particularly concerning the temperature level, it is important to identify those sensor locations that are suitable for the parameter identification and to calculate appropriate average values form multiple measurement devices.

In the following sections the relevant measurement sensors are introduced and their significance for the parameter estimation is discussed.

### 5.1.1
### Measurement of Cell Current and Cell Voltage

The electric measurements of cell current and cell voltage are very important to characterise the operating status of the fuel cell. They can be measured dynamically and with a high precision. For the estimation of the model parameters the direct current delivered by the stack is used because the subsequent transformation into rotating current is not considered here.

The cell voltage is measured at the complete stack, at groups of 49 cells and at selected single cells. In the course of the parameter estimation the average cell voltage of the whole stack is used and considered to be representative for a single cell.

### 5.1.2
### Temperature Measurement

The HotModule and the gas preparation unit are equipped with a number of thermocouples. Therefore sensors of type K are used[1] whose accuracy is given by $\pm 0.004$ T, but at least $\pm 2$ °C [1]. The position of the important thermocouples is shown in Fig. 5.1.

Here it can be seen that at the inlet of the HotModule the temperatures of air ($T_{air}$) and fuel gas ($T_{iir, in}$) are measured. They are needed as input conditions for the MCFC model and therefore have a high importance for the parameter identification. The thermocouple for the air temperature is located within the gas preparation unit behind the blower that draws in fresh air from the environment. This measurement is considered to be reliable.

At the fuel gas inlet the thermocouple is also installed within the gas preparation unit at the outlet of the external reformer. After passing the temperature measurement the fuel gas flows through an isolated and heated pipe of 3 m length before it enters the HotModule. According to the MTU the difference between the measured temperature $T_{iir, in}$ and the real temperature at the inlet of the HotModule is negligible because the heat losses are compensated by the heat tracing.

Within the HotModule the fuel gas first passes a heat exchanger where it is heated by the cathode gas. This heat exchanger is not accounted for in the model

---

1) SAB Type MTE302.

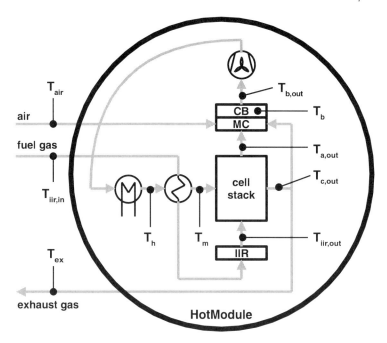

**Fig. 5.1** Position of the thermocouples within the HotModule
(MC – mixing chamber, CB – catalytic burner).

because it cannot be quantified by the measurement available. After the heat exchanger the fuel gas flows through the reforming units (IIR) and then into the gas manifold at the bottom of the HotModule. Here the temperature at the anode inlet is measured ($T_{iir, out}$). But within the whole manifold which has a size of approximately 2.5 m × 0.8 m there is only one single thermocouple installed. Hence the temperature that is measured here can only be used as a rough estimate because the temperature distribution within the manifold is by no means homogeneous.

At the outlet of the anode channels four thermocouples ($T_{a, out}$) are installed that are distributed over the stack area. With this configuration a more reliable measurement is possible than at the anode inlet. However, the difference between the individual sensors can be up to 20 K. For the parameter estimation a mean value of two thermocouples located in the middle of the stack is used because they are less influenced by fringe effects.

After passing the anode the gas goes into the mixing chamber and subsequently into the catalytic combustor. The temperature measurement here ($T_b$) is not needed for the parameter estimation. It is only used for a safety-related monitoring of the combustor. Afterwards the gas is forwarded to the cathode inlet by the two recycle blowers. At each blower a thermocouple is installed ($T_{b, out}$) but they do not deliver a representative value because the temperature difference between both measurements is up to 80 K. Thus they rather measure local temperatures than an average value.

At the cathode inlet an electric heater is installed that is mainly needed during the startup of the fuel cell. Within the normal operation that is investigated here the heater is switched off. Before the gas enters the cathode it flows through the heat exchanger which was already mentioned above. The temperature is measured both before $(T_h)$ and after $(T_m)$ the heat exchanger. For the parameter estimation the measurement $T_m$ directly at the cathode inlet is used. At this location six thermocouples are available distributed over the whole stack area. The average value of all sensors provides a reliable temperature at the cathode inlet.

Within the cell stack there are no temperature measurements available. Originally the stack was equipped with a number of thermocouples but due to the aggressive electrolyte nearly all the sensors failed during the beginning of the fuel cell operation.

The temperature of the exhaust air is measured at two points within the system. Firstly, there is a detailed measurement at the cathode $(T_{c,\,out})$ outlet similar to the cathode inlet with six thermocouples over the whole stack area. Secondly, the temperature of the exhaust air is measured at the outlet of the HotModule $(T_{ex})$. Both measurements show a very good agreement and the difference between the average of $T_{c,\,out}$ and $T_{ex}$ is only 1–2 K which is in the range of the accuracy of the thermocouples.

As explained before the thermocouples at the inlet and outlet of the cathode channels are distributed over the whole stack area as can be seen in Fig. 5.2. Unfortunately the number of sensors is too low to get a reliable information about the spatial temperature distribution that can be helpful for the parameter estimation. Therefore only the average values are used.

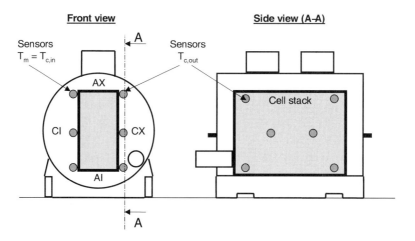

**Fig. 5.2** Configuration of the temperature sensors at the cathode inlet and outlet of the HotModule.

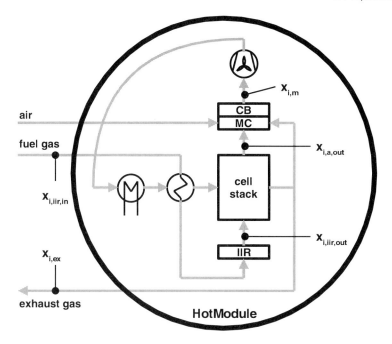

**Fig. 5.3** Position of the concentration measurements within the HotModule (MC – mixing chamber, CB – catalytic burner).

### 5.1.3
### Measurement of Concentrations

To determine the gas compositions at different points within the fuel cell system the HotModule is equipped with a gas chromatograph.[2] Figure 5.3 shows the position of the GC measurements. There is one measurement each at the fuel gas inlet ($x_{iir, in}$), at the anode inlet ($x_{iir, out}$) and at the exhaust air outlet ($x_{ex}$) respectively whereas there are four measurements at the anode outlet an two after the catalytic combustor ($x_m$).

The gas samples are drawn by small pipes that are connected to a multiposition valve. As the GC can only measure one sample at a time the nine positions are sequentially addressed by the valve. Therefore a complete measuring cycle takes approximately 60 min. So obviously the GC is suitable just for stationary measurements and dynamic changes can only be investigated with repeated measurements at a single position.

After passing the multiposition valve the sample is dried in a condenser because water must not get onto the GC column. Consequently the GC measures the composition of the dried gas while the water content is not determined.

---

**2)** Varian Micro-GC Bench model CP-2002.

Therefore a balancing of the HotModule is necessary as will be explained in Section 5.2.2.

In the course of the measurement campaigns the GC demanded a lot of attention. For instance the leak tightness of all connections had to be ensured and during the measurements the GC and the ambient conditions (e.g. temperature in the measuring cabinet) had to be monitored. Despite all efforts the GC did not show a satisfactory accuracy. Depending on the measuring position the total concentration was between 93 and 107% by volume. As this deviation cannot be assigned to a certain component the concentrations are scaled to 100%.

### 5.1.4
### Measurement of Flow Rates

Flow rates are only measured within the gas preparation unit outside the HotModule itself. There are three sensors that determine the amount of natural gas, water and air supplied to the fuel cell system. Their location is shown in Fig. 5.4. That means the flow rates of the fuel gas and exhaust gas as well as the flow rates within the Hotmodule are not available. They can only be calculated by a balancing of the system as explained in Section 5.2.2.

The flow rate of natural gas is measured between the activated carbon filter and the humidifier. This provides the amount of unreformed and dry gas. As measuring device a Coriolis Mass Flowmeter[3] is used with an accuracy of $\pm 0.5\%$ from the measured value [2]. For the measurement of the fresh water feed a high precision turbine wheel flowmeter[4] is used. It is installed after the water treatment and has an accuracy of $\pm 1.25\%$ regarding the full-scale value [3] which corresponds to $\pm 0.054$ l/min for the specific device. The flow rate of the supplied air is measured by a flow sensor that uses a thermal measuring technology.[5] The accuracy of the sensor is stated as $\pm 3.0\%$ of the measured value plus 0.4% of the full-scale value but at least 0.04 m/s [4].

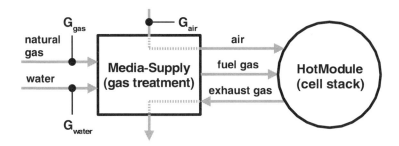

**Fig. 5.4** Position of flow rate measurements within the HotModule.

**3)** Endress+Hauser Proline promass 80F.
**4)** Kobold flowmeter PEL-L24SN0.
**5)** Schmidt Strömungssensor SS 20.60.

The reliable measurement of the mentioned flow rates is an important requirement for the parameter estimation because they determine the input of mass and energy into the system significantly.

## 5.1.5
### Conversion of the Measurements into Dimensionless Values

Because the MCFC model is written in dimensionless form all measurements have to be converted into dimensionless values before they can be used for the parameter estimation. A complete explanation of all dimensionless variables can be found in the thesis of Heidebrecht [5]. Only those correlations are given here that are relevant for the actual measurements.

The mole fractions are already in dimensionless form and can be directly used in the model equations. All other dimensionless values can be calculated by dividing the dimensional measured value by a particular standard value:

$$\vartheta = \frac{T}{T^0} \quad \text{with } T^0 = 298.15 \text{ K} \tag{5.1}$$

$$\Gamma = \frac{G}{G^0} \quad \text{with } G^0 = 6.377 \text{ mmol/s} \tag{5.2}$$

$$I_{\text{cell}} = \frac{\tilde{I}_{\text{cell}}}{I^0} \quad \text{with } I^0 = 1406.6 \text{ A} \tag{5.3}$$

$$U_{\text{cell}} = \frac{\tilde{U}_{\text{cell}}}{\phi^0} \quad \text{with } \phi^0 = 0.0257 \text{ V} \tag{5.4}$$

$$\tau = \frac{t}{t^0} \quad \text{with } t^0 = 12.15 \text{ s} \tag{5.5}$$

## 5.1.6
### Measurement Errors

The nominal accuracy of some sensors has already been given in the previous sections as they can be found in the data sheets of the manufacturers. But the reliability of the measurements at the HotModule is mainly influenced by systematic errors that have a much higher influence on the overall accuracy than the random errors particularly as the measuring noise is already reduced by the process control. The systematic errors mainly result from unfavourable positions of the sensors and the necessity to derive representative values for a single cell from the measurements at stack consiting of 342 cells.

The estimated measurement errors for all relevant measurement values are outlined in Table 5.1. The cell voltage can be measured very precisely for the whole stack with an accuracy of the sensor below ±1 mV. To get a representative single cell voltage the stack voltage is divided by the number of cells. Unfortu-

**Table 5.1** Expected measurement accuracy at the HotModule.

| Measurement | Dimensional | | Dimensionless | |
|---|---|---|---|---|
| | Symbol | Accuracy | Symbol | Accuracy |
| Cell voltage | $\tilde{U}_{\text{cell}}$ | $\pm 5$ mV | $U_{\text{cell}}$ | $\pm 0.1946$ |
| Temperature | $T_{\text{air}}$ | $\pm 3$ K | $\vartheta_{\text{air}}$ | $\pm 0.01$ |
| | $T_{\text{iir, in}}$ | $\pm 3$ K | $\vartheta_{\text{iir, in}}$ | $\pm 0.01$ |
| | $T_{\text{ex}}$ | $\pm 3$ K | $\vartheta_{\text{ex}}$ | $\pm 0.01$ |
| | $T_{\text{iir, out}}$ | $\pm 30$ K | $\vartheta_{\text{iir, out}}$ | $\pm 0.1$ |
| | $T_{a, \text{out}}$ | $\pm 20$ K | $\vartheta_{a, \text{out}}$ | $\pm 0.067$ |
| | $T_m$ | $\pm 10$ K | $\vartheta_m$ | $\pm 0.034$ |
| | $T_{c, \text{out}}$ | $\pm 10$ K | $\vartheta_{c, \text{out}}$ | $\pm 0.034$ |
| Concentration | $x_{i, \text{iir, in}}$ | $\pm 0.02$ | $\chi_{i, \text{iir, in}}$ | $\pm 0.02$ |
| | $x_{i, \text{ex}}$ | $\pm 0.02$ | $\chi_{i, \text{ex}}$ | $\pm 0.02$ |
| | $x_{i, \text{iir, out}}$ | $\pm 0.05$ | $\chi_{i, \text{iir, out}}$ | $\pm 0.05$ |
| | $x_{i, a, \text{out}}$ | $\pm 0.05$ | $\chi_{i, a, \text{out}}$ | $\pm 0.05$ |
| | $x_{i, m}$ | $\pm 0.05$ | $\chi_{i, m}$ | $\pm 0.05$ |
| Molar flow | $G_{\text{gas}}$ | $\pm 0.5\%$ | $\Gamma_{\text{gas}}$ | $\pm 0.5\%$ |
| | $G_{\text{water}}$ | $\pm 0.054$ l/min | $\Gamma_{\text{water}}$ | $\pm 0.0228$ |
| | $G_{\text{air}}$ | $\pm 3\%$ | $\Gamma_{\text{air}}$ | $\pm 3\%$ |

nately the comparison of this average value with measured voltages of single cell and cell packages shows differences of up to 20 mV. So the error for the representative single cell voltage is assumed to be higher than just the tolerance of the sensor.

The error of the temperature measurements depends on the location of the sensors. For $T_{\text{air}}$, $T_{\text{iir, in}}$ and $T_{\text{ex}}$ the thermocouples are located in the inlet and outlet pipes of the HotModule (see Fig. 5.1). Here a perfectly mixed gas can be assumed so that the error is mainly influenced by the tolerance of the sensors.

Compared to that there is a much bigger measurement uncertainty at the anode inlet ($T_{\text{iir, out}}$). At this location there is as well just one thermocouple installed but within the anodic gas manifold there is definitely no homogeneous mixture and the spatial distribution of the temperature cannot be detected by a single measurement. So this sensor only provides a rough estimate. In comparison the temperature measurement at the anode outlet ($T_{a, \text{out}}$) is more reliable because here four sensors are available.

An even better measurement information can be obtained at the cathode inlet ($T_m$) and outlet ($T_{c, \text{out}}$), where at each location six thermocouples are installed. The average of these values is representative for a single cell and shows a much lower measurement error than the anode temperatures.

In principle, the same considerations that apply for the temperatures are also valid for the concentrations. But additionally the inaccuracy of the gas chromatograph plays a major role as well (see Section 5.1.3). The measured gas compositions of the fuel gas ($x_{iir, in}$) and the exhaust air ($x_{ex}$) are representative because the samples are taken from a well mixed pipe. All other measurements only deliver selective information about spatially distributed states and therefore have a higher measurement error.

Another uncertainty concerning the gas concentrations arises from the fact that the composition of the unreformed natural gas fed into the HotModule is given by the supplier and cannot be measured. Because these data are provided only from time to time the actual gas composition during the time of the measurement can differ from that. But according to the operating company the fluctuations within the gas distribution system in Magdeburg are very low.

For the measurement of flow rates very precise sensors are installed because they provide important information for the process control. The accuracy of these devices corresponds to the tolerance given by the supplier (see Section 5.1.3).

## 5.1.7
## Measuring Campaigns

The experimental work comprised the measurement of steady state operating points at 50 mA/cm², 60 mA/cm², 70 mA/cm² and 80 mA/cm²). Current densities lower than 50 mA/cm² are unusual for normal operation the fuel cell and would be accompanied with other influences that disturb the measurements.[6] In addition current densities higher than 80 mA/cm² cannot be achieved with the HotModule in Magdeburg due to technical restrictions on the part of the power electronics. The result of these limitations is the above-mentioned operating range. The chosen increment of 10 mA/cm² corresponds to a change in the cell current of 79 A and an increase in the power of 15–20 kW. This allows a good discrimination between the individual operating points.

To ensure that the system is in steady state the measurements were carried out at the earliest one day after a load change. After verifying that the cell voltage and temperatures show a constant run the measurements were recorded over a period of several hours.

The investigation of the dynamic behaviour was also done in the range of 50 mA/cm² to 80 mA/cm² by applying load changes of 10 mA/cm² to the fuel cell. Because the concentrations cannot be measured during these dynamic processes the main focus of the experiments was to analyse the temperature dynamics.

---

**6)** For instance the electric heater would be necessary to keep the cell stack at a desired temperature level.

## 5.2
### Strategy for Parameter Estimation

The measurement data from the previous chapter are now used to estimate the MCFC model parameters. The aim of the parameter identification is to minimise the deviations between the measurements and the model simulations within the considered operating range. This results in a very complex optimisation problem as the spatially discretised MCFC model equations have to be considered as constraints. They consist of a set of highly nonlinear and coupled differential equations and numerous algebraic equations.[7]

Due to the complexity of the problem the simultaneous estimation of all model parameters considering all measured operating points by one single optimisation is numerically not feasible. This is particularly the case as the low precision of some measurements brings an additional uncertainty. Therefore it seems to be more effective to estimate only single parameters or groups of parameters, possibly by considering just parts of the model.[8]

To carry out this strategy it is very useful to have a complete set of measurement data available that includes the flow rates, concentrations and temperatures between all parts of the MCFC. Furthermore this data set should also fulfil the conservation laws of mass and energy to avoid inconsistent results. This can be achieved by a block-by-block plant balancing for every operating point.

Based on the consistent measurements a complete set of model parameters can be determined for every operating point. This leads to several sets of parameters depending on the operating conditions. Due to that a final optimisation is conducted to calculate a single parameter set that is valid for the whole operating range.

### 5.2.1
### Determination of Relevant Parameters

The MCFC model contains many parameters but only a part of them is relevant for the parameter estimation by measurement data. Many parameter values result from physical or chemical conditions (e.g. stoichiometric coefficients), from thermodynamic data (e.g. enthalpies of formation) or they are given by definition (e.g. standard values). Detailed explainations about that can be found in the thesis of Heidebrecht [5]. Therewith 31 parameters remain in the first instance that will be investigated more closely in the following. They contain mainly capacities and kinetic parameters:

- Volumes: $V_{iir}$, $V_a$, $V_c$, $V_m$
- Charge capacities: $c_a$, $c_e$, $c_c$

7) With a typical spatial discretisation of 10 × 10 the model contains 2709 differential equations and several thousand algebraic equations.

8) For example, only the mixing chamber is considered for the optimisation of the heat transfer coefficient $St_m$.

- Heat capacity: $c_{p,s}$
- Damköhler numbers: $Da_{\text{iir,ref}1/2}$, $Da_{a,\text{ref}1/2}$, $Da_{\text{ox}1/2}$, $Da_{\text{red}}$
- Arrhenius numbers: $Arr_{\text{iir,ref}1/2}$, $Arr_{a,\text{ref}1/2}$, $Arr_{\text{ox}1/2}$, $Arr_{\text{red}}$
- Stanton numbers: $St_{\text{iirs}}$, $St_{\text{as}}$, $St_{\text{cs}}$, $St_m$, $Pe_s$
- Mass transport coefficients: $D_{i,\text{as}}$, $D_{i,\text{cs}}$, $\kappa_e$
- Cathode recycle ratio: $R_{\text{back}}$

The dynamic behaviour of the cell and the related time constants are determined by capacities of mass, charge and energy. The mass storages depend on the volumes of the gas channels of the anode, cathode and IIR. They can be estimated from the dimensions of the cell.

The capacity of the electric double layers mainly affects the dynamics of the charge accumulation within the double layers. Because these processes happen within milliseconds they cannot be investigated with the available measuring devices. Furthermore, the charging of the double layers is not important for the control of the HotModule because there are no possibilities of influencing these fast processes. Heidebrecht [5] suggested some values for the charge capacities that are small enough to ensure a fast dynamic and therefore they are also used here.

The heat capacity $c_{p,s}$ represents by far the largest time constant of the system and influences primarily the slow temperature change within the solid phase. This parameter has to be estimated by measurements of the temperature behaviour during load changes. For a steady state operation of the fuel cell plant the heat capacity, as well as all other capacitive parameters, do not play any role.

The Damköhler numbers represent dimensionless reaction rate constants. They are very important kinetic parameters and have to be determined by the measurements. They determine the cell voltage and the concentrations at the outlet of the IIR, anode and cathode channels. As these states can be measured a good estimation of the Damköhler numbers is possible. However, to determine the Arrhenius numbers detailed kinetic experiments at varying temperatures would be necessary. But such measurements are not possible at the HotModule because the system is operated within a narrow temperature range and the temperature measurements are not precise enough. Therefore the values of the Arrhenius numbers are taken from the literature [6, 7].

The Stanton numbers can be interpreted as dimensionless heat exchange coefficients and they have a significant influence on the temperature distribution within the fuel cell. Based on the temperature measurements at the outlet of the gas channels the Stanton numbers can be estimated.

The Peclet number influences the heat conduction within the solid phase therewith affects the temperature profile within the cell. An estimation of the Peclet number would be possible by comparing measured and simulated temperature profiles whereas the significant influence of the (electro)chemical reactions also has to be considered. Unfortunately, there are no thermocouples available within the cell stack to measure the temperature profile. That means the Peclet number cannot be determined at the HotModule and the estimation of Heidebrecht [5] is used instead.

As the measurements at the HotModule where carried out at a medium cell current, mass transport limitations do not play a major role. Therefore the mass transport coefficients have to be chosen in such a way that their influence on the cell voltage is low. Heidebrecht [5] already proposed some values that correspond to these considerations so they are also used here.

The cathode recycle ratio specifies the amount of cathode gas that is returned to the catalytic combustor. This parameter can be easily estimated by the following correlation:

$$R_{back} = \frac{\Gamma_{back}}{\Gamma_{c,out}} = \frac{\Gamma_{back}}{\Gamma_{back} + \Gamma_{ex}} \tag{5.6}$$

This requires the values of the molar fluxes $\Gamma_{back}$ and $\Gamma_{ex}$ within the HotModule. These values cannot be measured directly but they can be calculated by balance equations (see Section 5.2.2). With this the recycle ratio can be calculated explicitly.

After the parameter analysis 12 parameters remain that have to be determined by the measurement data:

- Reaction kinetics: $Da_{iir,ref1/2}$, $Da_{a,ref1/2}$, $Da_{ox1/2}$, $Da_{red}$
- Heat transfer kinetics: $St_m$, $St_{iirs}$, $St_{as}$, $St_{cs}$
- Thermal capacity: $c_{p,s}$

## 5.2.2
### Balancing of the Fuel Cell Plant

To identify the model parameters a complete set of measurement data is required. On the one hand, it is necessary to know the input conditions of the MCFC model that include temperature, flow rate and composition of the fuel gas and air as well as the cell current and the energy input by the recycle blowers. On the other hand, measurements of significant states are needed to compare them to the model simulations. In this regard the cell voltage, temperatures and concentrations at different points within the system are important.

Unfortunately the measurements at the HotModule do not deliver all the necessary data. Some important values of flow rates and concentrations are missing. As explained in Section 5.1.4 only the flow rates of natural gas, water and air are measured but there are no such sensors within the fuel cell system. So the flow rates of the fuel gas, anode inlet, anode exhaust, cathode inlet, cathode reflux and exhaust air are unknown.

Furthermore, the GC only provides the composition of the dried gas so that the water content of these six gas flows is also not known. From this altogether 11 unknown values result that are listed in Table 5.2.[9]

---

9) As the compositions of the exhaust air and the cathode reflux are equal, the water content is also the same:
$x_{H_2O,back} = x_{H_2O,ex}$.

**Table 5.2** Non-measurable values that are needed for the parameter estimation.

| Molar flow: | $\Gamma_{iir,in}, \Gamma_{iir,out}, \Gamma_{a,out}, \Gamma_m, \Gamma_{back}, \Gamma_{ex}$ |
| Water concentration: | $\chi_{H_2O,iir,in}, \chi_{H_2O,iir,out}, \chi_{H_2O,a,out}, \chi_{H_2O,m}, \chi_{H_2O,ex}$ |

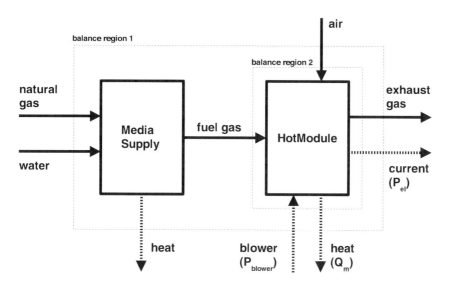

**Fig. 5.5** Balancing of the complete MCFC system with gas treatment and the HotModule.

To calculate these values a balancing of the mass and energy fluxes of the fuel cell system is conducted. For this purpose five suitable balance regions are defined within the gas treatment and the HotModule (see Figs. 5.5 and 5.6).

- Balance region 1 corresponds to the complete fuel cell plant consisting of the gas treatment and the HotModule. The inlet flows are natural gas, water and air. Furthermore the recycle blowers supply energy to the system. On the other hand, some of the cathode gas leaves the system as exhaust air and electric energy as well as heat are released to the environment.

- Balance region 2 refers to the housing of the HotModule. In comparison to balance region 1 the fuel cell is now fed with a pre-reformed and humidified fuel gas from the gas treatment.

  Within the HotModule some more balance regions are defined to calculate the internal mass and energy fluxes.

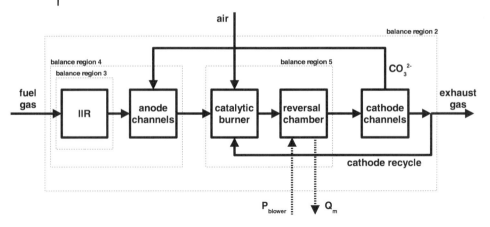

**Fig. 5.6** Balancing of the HotModule.

- Balance region 3 considers the reforming units within the cell stack where the indirect internal reforming (IIR) of the fuel gas takes place. After passing through the IIR the gas enters the anode channels.
- Balance region 4 is an extension of balance region 3 by the anode channels and the anode electrode. Here the boundary is between the anode and cathode side of the fuel cell in the middle of the electrolyte. Therefore the flow of carbonate ions ($CO_3^{2-}$) that goes through the carbonate melt from the cathode to the anode has to be balanced as well.
- Balance region 5 includes the catalytic combustor and the reversal chamber. In this case the cathode reflux is an important inlet flow besides the anode gas and air. The totally oxidised gas leaves the reversal chamber and is fed into the cathode channels. Besides that the energy input by the blowers and the heat losses to the environment have to be considered in the balances.

The balancing of the system is done for steady state conditions by considering every balance region as a black box so that only the inlet and outlet flows are taken into account. In doing so the particular processes within the balance region are not examined but it is demanded that the conservation laws of mass and energy are fulfilled. The aim of the fuel cell balancing is a set of equations to calculate the unknown values $u_k$ by using the measurements $y_h$:

$$f_b(y_h, u_k) = 0 \tag{5.7}$$

In the following the structure of the material and energy balances is explained.

## Material Balances

The material balances of the fuel cell system are set up using atom balances. All components within the system are formed by the elements N, C, O and H.[10] That means for every balance region four equations can be set up so that we get twenty material balance equations in total for all five balance regions.

As an example the H-balance for balance region 2 is shown in Eq. 5.8. The right-hand side of the equation describes the supply of hydrogen atoms by the fuel gas or more precisely by the components $CH_4$, $H_2$ and $H_2O$ in the gas. Because the supplied air is supposed to be dry it does not contain any hydrogen and is therefore not considered here. On the left-hand side of the equation the hydrogen atoms are accounted for that leave the balance region in the form of water with the exhaust air:

$$2\Gamma_{ex}^{tr} \frac{\chi_{H_2O, ex}}{1 - \chi_{H_2O, ex}} = \Gamma_{iir, in}^{tr} \left( 4\chi_{CH_4, iir, in}^{tr} + 2\chi_{H_2, iir, in}^{tr} + 2\frac{\chi_{H_2O, iir, in}}{1 - \chi_{H_2O, iir, in}} \right) \qquad (5.8)$$

This equation can be written in an implicit form (see Eq. (5.7)) where the measured values and the unknown values are clearly marked:

$$f_2(\underbrace{\chi_{CH_4, iir, in}^{tr}, \chi_{H_2, iir, in}^{tr}}_{\text{measured } y_h}, \underbrace{\Gamma_{ex}^{tr}, \chi_{H_2O, ex}, \Gamma_{iir, in}^{tr}, \chi_{H_2O, iir, in}}_{\text{unknown values } u_k}) = 0 \qquad (5.9)$$

All balance equations are written in terms of dry mole fractions so that the measurements from the gas chromatograph can be inserted directly. For instance in Eq. (5.8) this accounts for the concentrations of methane ($x_{CH_4, iir, in}^{tr}$) and hydrogen ($x_{H_2, iir, in}^{tr}$) in the fuel gas. A conversion between dry and wet mole fractions and molar fluxes respectively can be easily done:

$$\chi_i^{tr} = \frac{\chi_i}{1 - \chi_{H_2O}} \qquad \Gamma^{tr} = \Gamma \cdot (1 - \chi_{H_2O}) \qquad (5.10)$$

## Energy Balances

Complementary to the material flows the energy flows are also balanced at the fuel cell plant. But the energy balances can only be set up for balance regions 2 and 5 because for the other balance regions not all energy fluxes can be quantified, e.g. the heat losses by the media supply cannot be determined.

The formulation of the energy balances is done in the form of enthalpies. In doing so an isobaric system can be assumed because the investigated fuel cell only shows very small pressure differences. Furthermore, kinetic effects as well as the work performed by the volume forces can be neglected. With this the enthalpy balances for balance regions 2 and 5 read

---

10) The material flowing within the HotModule contain the components $CH_4$, $H_2O$, $H_2$, CO, $CO_2$, $O_2$ and $N_2$. Furthermore, we have $C_2H_6$, $C_3H_8$, $C_4H_{10}$, $C_5H_{12}$ and $C_6H_{14}$ in the natural gas.

$$h_{\text{iir, in}} + h_{\text{air}} + P_{\text{blower}} = h_{\text{ex}} + P_{\text{el}} + Q_m \qquad (5.11)$$

$$h_{a,\text{out}} + h_{\text{air}} + h_{\text{back}} + P_{\text{blower}} = h_m + Q_m \qquad (5.12)$$

The energy input into each balance volume is characterised by the enthalpy of convective mass fluxes $(H_k)$ and by the energy input of the recycle blowers $(P_{\text{blower}})$. The energy output also happens by convection but additionally by the heat losses to the environment $(Q_m)$. For balance region 2 the withdrawal of electric energy $(P_{\text{el}})$ has to be considered as well.

The enthalpy of each material flux can be calculated as the sum the enthalpy of its components (see Eq. (5.13)). The calculation of the temperature-dependent enthalpies of formation $h_{f,i}$ is done in the same way as in the MCFC model (see Chapter 3):

$$h = \Gamma \cdot \sum_i \chi_i \cdot h_{f,i}(\vartheta) \qquad (5.13)$$

The energy input by the recycle blowers is the product of the efficiency and the nominal power of the blowers at the given speed:

$$P_{\text{blower}} = \eta_{\text{blower}} \cdot P_{\text{blower}}^{\text{max}} \qquad (5.14)$$

The electric energy can be directly determined from the measurements of the cell current and cell voltage:

$$P_{\text{el}} = \frac{U_{\text{cell}} \cdot I_{\text{cell}}}{F} \qquad (5.15)$$

The heat losses at the HotModule are calculated as the product of the heat transfer coefficient $St_m$ and the temperature difference of the reversal chamber and ambient air:

$$Q_m = St_m \cdot (\vartheta_m - \vartheta_u) \qquad (5.16)$$

Detailed explanations about the estimation of $St_m$ can be found in Section 5.2.3.

**Minimisation of Measurement Errors**

The material and energy balances of the HotModule provide a set of 22 equations as listed in Table 5.3. They will be used to calculate the 11 unknown values that are outlined in Table 5.2.

With that the number of equations is larger than the number of unknowns, which means that the equation system is clearly *overdetermined*. In an ideal case, where the measurement data would be consistent, any selection of 11 linearly independent equations would be sufficient to solve the problem. But as explained in Section 5.1.6 the measurements contain uncertainties so that the

**Table 5.3** Available equations that result from mass and energy balances.

| | |
|---|---|
| Balance region 1 | four atom balances |
| Balance region 2 | four atom balances |
| | one enthalpy balance |
| Balance region 3 | four atom balances |
| Balance region 4 | four atom balances |
| Balance region 5 | four atom balances |
| | one enthalpy balance |
| Total | 22 balance equations |

result for $u_k$ is much different depending on the chosen equations. Furthermore, it is not possible to identify 11 balance equations that give the 'best' solution because in either case some other balances are violated. That means the measurements are not consistent regarding the conservation laws of mass and energy because they contain measurement errors.

As the MCFC model is strictly based on these balance equations it is also required that the measurement data used for the parameter identification fulfil the laws of mass and energy conservation. Therefore, the measurement uncertainties are included in the calculations. To achieve this corrected measurement values $\tilde{y}_h$ are introduced which shall exactly comply with the balance equations. The difference between the measured value $y_h$ and the corrected value $\tilde{y}_h$ is the assumed measurement error of a particular measurement. As the measurement errors and therewith also the corrected values are unknown they have to be calculated by the balance equations as well. That means in addition to the primarily unknown value $u_k$ another 29 corrected measurement values $\tilde{y}_h$ have to be considered. They consist of 21 concentrations, 5 temperatures as well as the material flows of natural gas, water and air as listed in Table 5.4.

The number of equations also increases because five closing conditions have to be considered for the mole fractions:

$$\sum_i \tilde{\chi}^{tr}_{i,\text{iir,in}} = \sum_i \tilde{\chi}^{tr}_{i,\text{iir,out}} = \sum_i \tilde{\chi}^{tr}_{i,a,\text{out}} = \sum_i \tilde{\chi}^{tr}_{i,m} = \sum_i \tilde{\chi}^{tr}_{i,\text{ex}} = 1 \qquad (5.17)$$

**Table 5.4** Overview of all measurement data used for the fuel cell balancing.

| | |
|---|---|
| Molar flows | $\Gamma_{\text{gas}}, \Gamma_{\text{water}}, \Gamma_{\text{air}}$ |
| Temperatures | $\vartheta_{\text{air}}, \vartheta_{\text{iir,in}}, \vartheta_{a,\text{out}}, \vartheta_m, \vartheta_{\text{ex}}$ |
| Concentrations | $\chi^{tr}_{i,\text{iir,in}}, \chi^{tr}_{i,\text{iir,out}}, \chi^{tr}_{i,a,\text{out}}$ with $i \in \{CH_4, H_2, CO, CO_2, N_2\}$ |
| | $\chi^{tr}_{i,m}, \chi^{tr}_{i,\text{ex}}$ with $i \in \{CO_2, O_2, N_2\}$ |

**Table 5.5** Weighting factors used for the error minimisation.

| Measured value | Weighting factor $1/w_h$ |
|---|---|
| $\Gamma_{gas}$ | $0.005 \times \Gamma_{gas}$ |
| $\Gamma_{water}$ | $0.0228$ |
| $\Gamma_{air}$ | $0.03 \times \Gamma_{air}$ |
| $\chi_{i,\,iir,\,in}^{tr},\ \chi_{i,\,a,\,out}^{tr},\ \chi_{i,\,m}^{tr},\ \chi_{i,\,ex}^{tr}$ | $0.1 \cdot \chi_{i,\,s}^{tr}$ |
| $\chi_{i,\,iir,\,out}^{tr}$ | $0.2 \times \chi_{i,\,iir,\,out}^{tr}$ |
| $\vartheta_{iir,\,in},\ \vartheta_{air},\ \vartheta_{ex}$ | $0.01$ |
| $\vartheta_{a,\,out},\ \vartheta_{m}$ | $0.1$ |

Hence the introduction of corrected measurement values leads to a set of 27 equations[11] that contain 40 unknown values. With this the equation system is now 13-fold *under*determined. Thus a unique solution is still not possible but the problem can now be solved by an optimisation.

The aim is to determine a set of corrected measurement values $\tilde{y}_h$ and unknown values $u_k$ that show a minimal deviation between $y_h$ and $\tilde{y}_h$ and also fulfil the conservation laws. From this requirement the objective function of the optimisation problem can be written as a sum of squares:

$$F(\tilde{y}_h, u_k) = \sum_h (w_h \cdot (\tilde{y}_h - y_h))^2 \overset{!}{\to} \min \tag{5.18}$$

s.t.

$$f_b(\tilde{y}_h, u_k) = 0 \tag{5.19}$$

$$y_h^{\min} \leq \tilde{y}_h \leq y_h^{\max} \tag{5.20}$$

The optimisation variables of this problem are all unknown values $u_k$ from Table 5.2 as well as the corrected measurement values $\tilde{y}_h$ from Table 5.4. Each term of the objective function is weighted by a factor $w_j$ which accounts for the reliability of every measured value. If the expected measurement error is relatively high the weighting factor of this value is low, which means that a correction of this measurement has a lower influence. A complete list of all weighting factors can be found in Table 5.5. They correspond to the expected measurement errors in Table 5.1. The mole fractions have the smallest weighting factor because of the high measurement uncertainty. The temperatures have a stronger influence depending on the position and number of sensors. As the measurements of the mass fluxes are very accurate they have the highest weighting factors.

---

11) 20 material balances, 2 enthalpy balances, 5 closing conditions.

The inequality constraints in Eq. (5.20) define a range for every measured value where the corrected values have to stay during the optimisation. The upper and lower boundaries reflect the accuracy of the measurement and account for technical restrictions of the fuel cell system.

The optimisation problem is solved with a SQP algorithm in MATLAB using the function fmincon. Therefore, suitable initial values are required. For the unknown variables $u_k$ they can be determined by a selection of 11 linearly independent material balances. As initial values for the corrected measurement values $\tilde{y}_h$ the original measurements are used.

It is possible to put the optimisation problem in a more compact form by setting up the balance equations in terms of component-based molar fluxes. To obtain this the mole fractions in the balance equations are replaced by their definition:

$$\chi_i^{tr} = \frac{\Gamma_i}{\Gamma^{tr}} \qquad \chi_{H_2O} = \frac{\Gamma_{H_2O}}{\Gamma^{tr} + \Gamma_{H_2O}} \qquad \Gamma^{tr} = \sum_{i \neq H_2O} \Gamma_i \tag{5.21}$$

As an example the hydrogen balance for balance region 2 (see Eq. (5.8)) is simplified as follows:

$$4 \cdot \Gamma_{CH_4, iir, in} + 2 \cdot \Gamma_{H_2, iir, in} + 2 \cdot \Gamma_{H_2O, iir, in} - 2 \cdot \Gamma_{H_2O, ex} = 0 \tag{5.22}$$

In total the number of optimisation variables is reduced to 35 by this procedure. At the same time the closing conditions for the mole fractions are dropped so that only 22 balance equations have to be considered now. But this does not change anything about the nature of the problem because the equation system is still 13-fold *under*determined and the optimisation has to be carried out as explained before.

## 5.2.3
### Sensitivity Analysis

The following sensitivity analysis investigates the influence of these relevant parameters on the fuel cell states. Thereby the heat capacity of the solid phase, $c_{p,s}$, is not included in the first instance because it only influences the dynamic behaviour and does not effect the steady state. Then, after estimating the kinetic parameters from steady state measurement also the heat capacity will be determined by analysing the dynamic system behaviour.

Thus the main focus of the parameter estimation lies on the Damköhler and Stanton numbers. From the literature (e.g. [8]) it is well known that the steam reforming within the MCFC is very fast. To draw a conclusion regarding the Damköhler numbers $Da_{iir, ref1}$ and $Da_{iir, ref2}$ of the reforming reaction within the IIR the gas composition at the outlet of the IIR is analysed. By doing so it can be concluded that the water–gas shift reaction (ref2) is actually close to equilibrium for all load cases. Furthermore, it seems that based on the temperature and gas composition at the IIR outlet the steam reforming is even beyond the equilibrium. On the one hand, this is due to high measurement uncertainties in the anodic gas manifold. But on the other hand, a more important effect is the inflow of air into the manifold that changes the gas composition. For instance by supplying air to the anode gas the methane is oxidised and it seems as if this had been converted into the IIR.

Based on these observations it can be at least assumed that the reforming reactions in the IIR are very fast and close to equilibrium. Therefore, sufficiently high values are assigned to the Damköhler numbers:

$$Da_{iir, ref1} = Da_{iir, ref2} = 200 \tag{5.23}$$

As the same catalyst materials are used for the IIR and the direct internal reforming (DIR) it can also be assumed that the DIR reactions in the anode are also very fast. According to this the Damköhler numbers $Da_{a, ref1}$ and $Da_{a, ref2}$ are set to equivalent values so that reforming reactions are close to equilibrium:

$$Da_{a, ref1} = Da_{a, ref2} = 200 \tag{5.24}$$

As Heidebrecht [5] already stated in his thesis the oxidation of carbon monoxide only has a little influence on the fuel cell states. In particular this can be clearly seen at the current–voltage curve in Fig. 5.7. Over a wide operating range the cell voltage is nearly identical independent from the CO oxidation. Only at high current voltages the curves show a slightly different shape and the influence

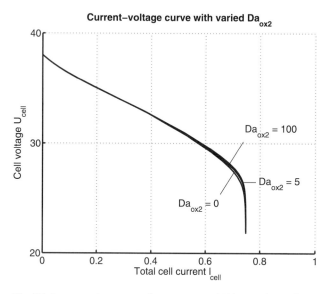

**Fig. 5.7** Current–voltage curve for various Damköhler numbers of the CO oxidation.

of $Da_{ox2}$ can be seen. As the operating range of the HotModule in Magdeburg is far below the limiting current the oxidation of carbon monoxide does not play an important role and can be neglected:

$$Da_{ox2} = 0 \tag{5.25}$$

Thus the remaining Damköhler numbers are $Da_{ox1}$ at the anode and $Da_{red}$ at the cathode. They determine the rates of the electrochemical reactions or at given reaction rates (cell current) the cell voltage respectively.

The Stanton number $St_m$ is the dimensionless heat transfer coefficient that influences the heat transfer between reversal chamber and the environment. For the estimation of $St_m$ the enthalpy balance of the reversal chamber can be used. This reads as follows:

$$0 = \Gamma_{b,\,\mathrm{out}}(\vartheta_{b,\,\mathrm{out}} - \vartheta_m) \sum_i \chi_{i,\,b,\,\mathrm{out}} c_{p,\,i} - Q_m - P_{\mathrm{blower}} \tag{5.26}$$

Together with the equation for the heat losses

$$Q_m = St_m(\vartheta_m - \vartheta_u) \tag{5.27}$$

a relation for $St_m$ can be derived:

$$St_m = \frac{\Gamma_{b,\,\mathrm{out}}(\vartheta_{b,\,\mathrm{out}} - \vartheta_m) \sum_i \chi_{i,\,b,\,\mathrm{out}} c_{p,\,i} - P_{\mathrm{blower}}}{\vartheta_m - \vartheta_u} \tag{5.28}$$

The values for $\Gamma_{b,\,\mathrm{out}}$, $\vartheta_m$ and $\vartheta_u$ in this equation can be determined from the measurements and the balance equations with good accuracy. Also the average heat capacity $c_p$ of the gas can be calculated from the measured gas composition and temperature. Only the temperature $\vartheta_{b,\,\mathrm{out}}$ at the outlet of the catalytic combustor cannot be determined reliably. As explained in Section 5.1.2 the thermocouples at the recycle blowers show a difference of up to 80 K. Depending on which temperature measurement is used the resulting Stanton number range from $St_m = -7.0, \ldots, 4.0$, whereas negative values are even unphysical. So obviously the estimation of $St_m$ is not possible in this way.

Another possibility of determining $St_m$ is by the estimation of the heat losses $Q_m$ over the outer vessel of the HotModule. Therefore the heat transfer between the HotModule and the environment caused by convection and radiation is determined. The result for the emitted power is $P_{\mathrm{heat}} = 5.8$ kW which corresponds to a dimensionless heat flux of $Q_m \approx 1$. Considering a temperature inside the HotModule of 600–650 °C ($\vartheta_m \approx 3$) and a ambient temperature of 25 °C ($\vartheta_u \approx 1$) an estimation for $St_m$ is given by Eq. (5.27):

$$St_m = \frac{Q_m}{\vartheta_m - \vartheta_u} \approx \frac{1}{3 - 1} = 0.5 \tag{5.29}$$

With this the number of parameters that still have to be determined is reduced to 5 namely the Damköhler numbers $Da_{ox1}$ and $Da_{red}$ as well as the Stanton numbers $St_{iirs}$, $St_{as}$ and $St_{cs}$. Each of these parameters influences indirectly more or less all states of the fuel cell but some values respond much stronger to parameter changes than others. To investigate the influence that the remaining parameters have on the important system states a sensitivity analysis of the steady state MCFC model is carried out. This gives information about the measurement data that can be used to estimate the different parameters. The following states are included in the sensitivity analysis:

- Temperature $\vartheta_{iir,out}$ and concentrations $x_{i,iir,out}$ at the anode inlet
- Temperature $\vartheta_{a,out}$ and concentrations $x_{i,a,out}$ at the anode outlet
- Temperature $\vartheta_m$ at the cathode inlet
- Temperature $\vartheta_{c,out}$ at the cathode outlet
- Cell voltage $U_{cell}$

Because the gas composition at the cathode inlet and outlet is determined by simple combustion calculations and therefore independent of any kinetic system parameters they are not considered here. Table 5.6 shows the qualitative result of the sensitivity analysis. If this table is read line by line it shows which parameter influences the particular value most. The marking '++' means that the respective state is very sensitive to a certain parameter while '+' shows that the influence of this parameter is bigger than for other parameters.

It is obvious that $Da_{red}$, which is the dimensionless reaction rate constant of the reduction reaction at the cathode, has a big influence on all important states. It mainly determines the cell voltage and thus the withdrawal of electric energy from the system. As this energy would otherwise accumulate as thermal energy the cell voltage also influences the cell temperature and via the chemical equilibrium also the molar fractions within the gas channels. So the Damköhler number $Da_{red}$ is one of the most important parameters of the model. Furthermore, a clear

**Table 5.6** Influence of the model parameters on the most important system states.

| | $St_{iirs}$ | $St_{as}$ | $St_{cs}$ | $Da_{ox1}$ | $Da_{red}$ |
|---|---|---|---|---|---|
| $\vartheta_{iir,out}$ | ++ | | | | + |
| $x_{i,iir,out}$ | ++ | | | | + |
| $\vartheta_{a,out}$ | | + | | | + |
| $\vartheta_m$ | | | | | + |
| $\vartheta_{c,out}$ | | | | | + |
| $x_{i,a,out}$ | + | + | | | ++ |
| $U_{cell}$ | | | | + | ++ |

sensitivity of the cell voltage $U_{cell}$ to $Da_{ox1}$ can be observed, whereas the influence of $Da_{red}$ is still bigger.

Another major parameter is the Stanton number $St_{iirs}$ which determines the heat transfer from the solid phase to the IIR. The heat input into the reforming units again influences the strongly endothermal reforming reactions. Thus the gas composition and temperature at the outlet of the IIR, which is equal to the anode inlet, is strongly sensitive to $St_{iirs}$. The two other Stanton numbers, $St_{as}$ and $St_{cs}$, mainly influence the temperatures within the gas channels, however $St_{cs}$ has no significant influence on the measured values.

If the heat losses to the environment are assumed to be constant the temperature $\vartheta_{c,out}$ of the exhaust air only depends on the electric power for a given fuel gas and air supply. For a constant cell current the electric power in turn only depends on the cell voltage which is mainly influenced by $Da_{red}$. In contrast this means that the measurements for $\vartheta_{c,out}$ do not have to be considered during the parameter estimation. If the parameters, especially $Da_{red}$, are chosen in such a way that the simulated cell voltage shows a good agreement with the measurements it also means that the cathode temperature will also fit very well.

## 5.2.4
### Parameter Estimation for a Single Load Case

The result of fuel cell balancing is a complete set of measurement data for every investigated load case. Furthermore, five parameters could be identified by a sensitivity analysis that have to be estimated with the help of experimental data. The next step is an iterative procedure to estimate the model parameters for every operating point. To quantify how good the simulated system states $\hat{y}_h$ correspond to the (corrected) measurement data $\tilde{y}_h$ a sum of squared errors is used. The aim of the parameter estimation is to minimise this objective function:

$$F_r(p_m) = \sum_h (v_h \cdot (\tilde{y}_h - \hat{y}_h(p_m)))^2 \overset{!}{\to} \min \tag{5.30}$$

In doing so each term is weighted by a factor $v_h$ which determines the effect of the particular states on the objective function. The factors are chosen in such a way that based on the same relative deviation between measurement and simulation the cell voltage has a bigger influence on the objective function than the temperatures which again weigh more than the concentrations (see Table 5.7). This means that a deviation of the cell voltage is less tolerated and the optimisation algorithm tries to fit this value as good as possible. This is justified by the fact that the cell voltage is the most important state and furthermore can be measured very precisely. The temperatures are also important states but they cannot be measured with such a high accuracy and therefore a larger deviation is tolerated here. The same applies for the gas concentrations that are measured with the lowest precision. The solution of the optimisation problem in Eq. (5.30) is much

**Table 5.7** Order of magnitude of the important states and their weighting in the objective function.

| State | Order of magnitude | Weighting factor |
|---|---|---|
| Concentrations | $\chi = 0, \ldots, 1$ | $v_\chi = 1$ |
| Temperatures | $\vartheta \approx 3$ | $v_\vartheta = 10/3$ |
| Cell voltage | $U_{\text{cell}} \approx 30$ | $v_U = 10/3$ |

more complex than it might seem at first sight. This is due to the fact that the simulated states $\hat{y}_h$ are calculated by the MCFC model from Chapter 3, which consists of a set of highly nonlinear and coupled differential equations. They represent the equality constraints of the optimisation and make the numerical solution very difficult. Because of this the simultaneous estimation of all parameters $p_m$ is only possible with good starting values. As they are not available in the beginning the sensitivities from Table 5.6 are used to determine the parameters in an iterative procedure.

The sensitivity analysis in Section 5.2.3 shows that certain parameters or groups of parameters have a strong influence on some measurable states while other values do not respond significantly on them. This situation is used to estimate single parameters or small groups of parameters in such a way that the states which they influence most correspond to the measurements.

At first the Damköhler number $St_{\text{iirs}}$ is varied while all other parameters remain constant. By changing the heat transfer between IIR and solid phase the temperature and gas composition at the outlet of the IIR ($=$ anode inlet) are adjusted so that they reflect the measurements as good as possible. Therefore, the objective function given in Eq. (5.30) is set up only for the relevant states of the IIR and minimised by changing $St_{\text{iirs}}$:

$$F_r^{\text{iir}}(St_{\text{iirs}}) = \sum_i \left( v_\chi \cdot \left( \tilde{\chi}_{i,\,\text{iir, out}} - \hat{\chi}_{i,\,\text{iir, out}}(St_{\text{iirs}}) \right) \right)^2$$

$$+ \left( v_\vartheta \cdot \left( \tilde{\vartheta}_{\text{iir, out}} - \hat{\vartheta}_{\text{iir, out}}(St_{\text{iirs}}) \right) \right)^2 \overset{!}{\to} \min \tag{5.31}$$

In a second step the two Stanton numbers $St_{\text{as}}$ and $St_{\text{cs}}$ are estimated so that the temperatures at the anode inlet and outlet correspond to the measured temperatures best possible. Thereby the previously estimated Stanton number $St_{\text{iirs}}$ is kept constant as well as the Damköhler numbers $Da_{\text{ox1}}$ and $Da_{\text{red}}$. In this case the objective function consists only of those terms that are sensitive regarding the two Stanton numbers:

$$F_r^{St}(St_{\text{as}}, St_{\text{cs}}) = \left( \tilde{\vartheta}_{\text{iir, out}} - \hat{\vartheta}_{\text{iir, out}}(St_{\text{as}}, St_{\text{cs}}) \right)^2$$

$$+ \left( \tilde{\vartheta}_{a,\,\text{out}} - \hat{\vartheta}_{a,\,\text{out}}(St_{\text{as}}, St_{\text{cs}}) \right)^2 \overset{!}{\to} \min \tag{5.32}$$

In the third step the Damköhler numbers $Da_{ox1}$ and $Da_{red}$ are optimised to adjust the cell voltage and the composition of the anode gas. Again all other parameters remain constant. The estimation of the Damköhler numbers is also done by minimising the corresponding objective function. It only contains the cell voltage and the anode outlet concentrations which are mainly influenced by the Damköhler numbers:

$$F_r^{Da}(Da_{ox1}, Da_{red})$$

$$= (v_U \cdot (\tilde{U}_{cell} - \hat{U}_{cell}(Da_{ox1}, Da_{red})))^2$$

$$+ \sum_i (v_\chi \cdot (\tilde{x}_{i,a,out} - \hat{x}_{i,a,out}(Da_{ox1}, Da_{red})))^2 \overset{!}{\to} \min \qquad (5.33)$$

These three optimisation steps are carried out consecutively until an optimum is achieved or sufficient staring values are available that enable a simultaneous adjustment of all parameters. This iterative procedure which optimises only parts of the objective function does not necessarily lead to a minimum of the original objective function (5.30). Only if the single terms of the objective function show strong sensitivities to the parameters this is ensured.

The result of the optimisations described in this chapter is a set of parameters that gives the best possible agreement between simulations and measurements for a single steady state operating point of the fuel cell plant.

## 5.2.5
### Parameter Estimation for the Whole Operating Range

The parameter estimation described in the previous chapter has to be performed for every measured operating point independently. This leads to a set of optimal parameters for every set of measurement data. But these parameters can vary. Because the aim is to find a single set of parameters that is valid for the whole operating range another optimisation has to be performed. For this purpose a new objective function $F_{total}$ is set up that contains the sums of squared errors for all operating conditions. Now the model parameters have to be adjusted so that the total objective function is minimised:

$$F_{total}(p_m) = \sum_r F_r(p_m) \overset{!}{\to} \min \qquad (5.34)$$

As the MCFC model is so complex that even the simultaneous parameter estimation for a single operating point requires a large numerical effort the optimisation of all operating points together cannot be done within reasonable time and effort. The optimisation of the total objective function would require the simultaneous solution of the MCFC model with different input conditions for each operating point which is not possible with the available resources. Therefore, the pa-

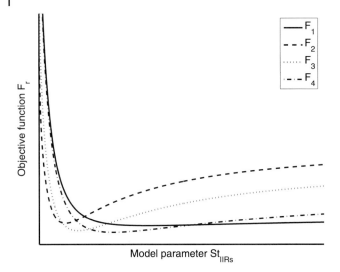

**Fig. 5.8** Qualitative illustration of the objective function for different operating conditions depending on the model parameter $St_{iirs}$.

rameters are determined in an iterative procedure as in the previous chapter. But here the objective function in Eq. 5.34 has to be minimised.

Therewith the estimation of $St_{iirs}$ leads to the following optimisation problem:

$$F_{total}(St_{iirs}) = \sum_r F_r(St_{iirs}) \overset{!}{\to} \min \tag{5.35}$$

Here the difficulty is that $F_{total}$ cannot be calculated explicitly. Therefore, at first the values for the single objective functions $F_r$ are calculated by varying $St_{iirs}$ as shown in Fig. 5.8. The sum of these objective functions gives a good approximation for the total objective function. Because $F_{total}$ is unimodal in the relevant interval the minimum of this function can be easily determined (Fig. 5.9).

The next step is the estimation of the Damköhler numbers $Da_{ox1}$ and $Da_{red}$. These two parameters are correlated so that they cannot be estimated separately in the same way as $St_{iirs}$. Instead they have to be determined simultaneously which leads to a two-dimensional optimisation:

$$F_{total}(Da_{ox1}, Da_{red}) = \sum_r F_r(Da_{ox1}, Da_{red}) \overset{!}{\to} \min \tag{5.36}$$

For an efficient solution of this problem an approach is chosen that uses an approximation of the objective function:

$$F_r(Da_{ox1}, Da_{red}) \approx G_r(Da_{ox1}, Da_{red}) \tag{5.37}$$

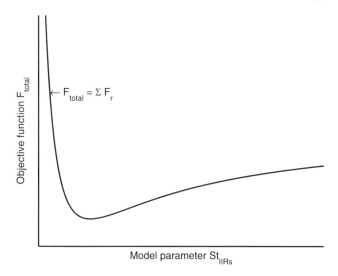

**Fig. 5.9** Qualitative illustration of the total objective function depending on the model parameter $St_{iirs}$.

Here the original objective function $F_r$ is substituted by a function $G_r$ that serves as a good approximate and can be calculated explicitly with a significantly less numerical effort. To ensure a good agreement between $F_r$ and $G_r$ an approximation is chosen depending on the shape of the objective function. Figures 5.10 and 5.11 exemplarily show $F_r$ depending on just one parameter. As can be seen here the

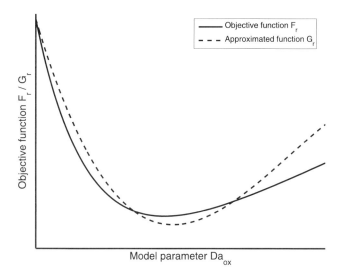

**Fig. 5.10** Qualitative illustration of the total objective function and its approximation depending on the model parameter $Da_{ox1}$.

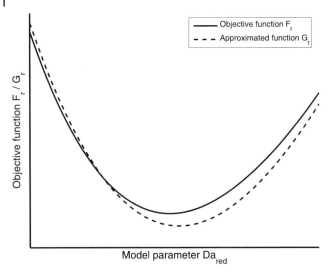

**Fig. 5.11** Qualitative illustration of the total objective function and its approximation depending on the model parameter $Da_{red}$.

curves have an asymmetric form so that a quadratic approximation would not be sufficient. To better reproduce the objective function a polynomial of third order is chosen in this case:

$$G_r(Da_{ox1}, Da_{red}) = \mathbf{a}_r \cdot (Da_{ox1})^3 + \mathbf{b}_r \cdot (Da_{red})^3$$
$$+ \mathbf{c}_r \cdot (Da_{ox1})^2 \cdot Da_{red} + \mathbf{d}_r \cdot Da_{ox1} \cdot (Da_{red})^2$$
$$+ \mathbf{e}_r \cdot (Da_{ox1})^2 + \mathbf{f}_r \cdot (Da_{red})^2 + \mathbf{g}_r \cdot Da_{ox1} \cdot Da_{red}$$
$$+ \mathbf{h}_r \cdot Da_{ox1} + \mathbf{i}_r \cdot Da_{red} + \mathbf{j}_r \qquad (5.38)$$

The polynomial $G_r$ contains 10 coefficients $\mathbf{a}_r, \ldots, \mathbf{j}_r$, which have to be calculated independently for every operating point. Therefore 10 nodes are required that have to be chosen in such a way that the objective function is approximated as good as possible near the particular minimum.[12] The selection of the nodes is illustrated in Fig. 5.12. At first, four parameter variations are carried out (dashed lines). With this numerous values of the objective function are available that can be used directly or to calculate partial derivatives (difference quotients). Thereby the kind and the position of the nodes are chosen to fit the objective function as good as possible (see Figs. 5.10 and 5.11).

By inserting the nodes into Eq. (5.38) or its partial derivative respectively a linear equation system is obtained that has to be solved in order to determine the

**12)** The location of the minima of the single operating points is already known from previous optimisations.

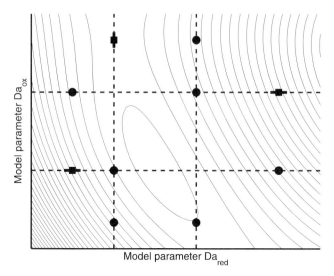

**Fig. 5.12** Qualitative illustration of the objective function as a contour plot. The nodes that were chosen for evaluating the approximated function are marked as circles (function values) and squares (partial derivatives along the indicated direction).

coefficients of $G_r$. But, a solution is only possible if the equations are linearly independent, which depends on the selection of the nodes. If the equation system cannot be solved the nodes have to be changed.

In this way an approximated objective function $G_r$ is calculated for every operating point. In the last step they all are inserted into Eq. (5.36):

$$F_{total}(Da_{ox1}, Da_{red}) \approx \sum_r G_r(Da_{ox1}, Da_{red}) \overset{!}{\to} \min \qquad (5.39)$$

This gives an approximation for the total objective function $F_{total}$ that can now be calculated explicitly. The following optimisation of the model parameters $Da_{ox1}$ and $Da_{red}$ can be done with a conventional SQP solver.

In the same way as explained for the Damköhler numbers the two Stanton numbers $St_{as}$ and $St_{cs}$ can be estimated with the help of approximated objective functions. Afterwards the iterative estimation of the parameter groups $St_{iirs}$, $Da_{ox1}/Da_{red}$ and $St_{as}/St_{cs}$ is done until the minimum of the total objective function is determined with sufficient precision.

### 5.2.6
### Temperature Dynamics

In the previous chapters all those parameters were estimated that have an influence on the steady state of the MCFC. Furthermore, there are capacitive parameters that determine the dynamic behaviour of the system as already discussed in Section 5.2.1. The most important parameter of this group is the heat capacity $c_{p,s}$ of the solid phase which is also the only parameter that can be estimated from measurements at the HotModule. The heat capacity represents by far the largest time constant of the system which is responsible for the very slow temperature dynamic of the solid parts of the fuel cell.

The estimation of $c_{p,s}$ is done by comparing measured and simulated changes of the cell voltage and temperatures during load changes. Then the heat capacity is chosen so that the simulated time until a new steady state is reached agrees with the measured time that the real MCFC system needs for a load change.

## 5.3
## Results of the Parameter Identification

In the previous two chapters the measurements at the HotModule and the strategy for the parameter estimation have been explained. At this point the results of this approach will be shown and discussed.

At first some measurement data from steady state operations are presented. Then the plant balancing and error minimisation are carried out. This results in several sets of measurement data that are used afterwards to determine most of the model parameters. Beyond that the estimation of the heat capacity of the solid phase requires the analysis of dynamic measurements.

### 5.3.1
### Steady State Measurements

The measurements at steady state operation were done at four different operation points characterised by their particular current density: $i_{cell} = 50$ mA/cm$^2$, 60 mA/cm$^2$, 70 mA/cm$^2$ and 80 mA/cm$^2$. Depending on that all other input conditions are set by the process control. The measurements were done after the fuel cell had been at a certain operating point for a few days to assure a steady state. Afterwards all continuously available measurements, which are cell current, cell voltage as well as temperatures and flow rates, were recorded for several hours. Furthermore, at the end of the measurement cycle all concentrations were measured.

In the following some important results of the steady state measurements are discussed. A diagram of the cell voltage is shown in Fig. 5.13. Here a significant noise of the measurement data can be observed. The cell voltage fluctuates by $\pm 5$ mV in terms of the mean value. Thus the difference between the single operating points becomes smaller, especially the cell voltages of 70 mA/cm$^2$ and 80 mA/cm$^2$ are close together.

As the accuracy of the sensor is well below 1 mV the fluctuations of the cell voltage are not the effect of stochastical measurement errors. It is rather caused by non-quantifiable disturbances that force the process control to adjust the input

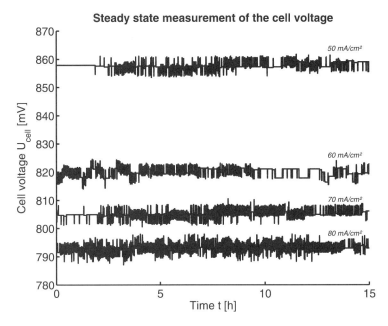

**Fig. 5.13** Measurement of the cell voltage at steady state for different current densities.

conditions. Even the cell current that is given as a constant value shows clear fluctuations of up to $\pm 15$ A (approximately $\pm 1$ mA/cm$^2$) around the setpoint as can be seen in Fig. 5.14. The graph of the natural gas supply, which is another important input parameter, is given in Fig. 5.15. Also in this diagram the influence of the process control can be seen very clearly. The fluctuation of the natural gas flow rate is up to $\pm 5\%$ of the average value, whereas the flow rate seems to oscillate between two discrete conditions.

As the natural gas flow rate determines the main energy input into the system it more or less influences all other states. So, besides the cell voltage, the measured temperatures fluctuate clearly as can be seen in Fig. 5.16 for the anode outlet temperature. But the shape of this curve cannot be explained just with fluctuating fuel gas flow rates. It is again a complex mix of environmental influences and the process control of the fuel cell system.

While comparing the temperature measurements to the other diagrams one can observe that the signal is less noisy. The reason for this is a different filtering of the signals. For every sensor a specific threshold level is defined that needs to be exceeded; otherwise the process control does not recognise a change of this value. For most temperatures the threshold is 1 K. This leads to a smoother curve and the anode temperature does not seem to fluctuate so much. But the maximum oscillations over the whole measuring period are not affected by this. For the anode outlet temperature they are up to $\pm 5$ K in terms of the average value.

For the parameter identification the average value of each measured value for the whole measuring period is used. Obviously the accuracy of the values that

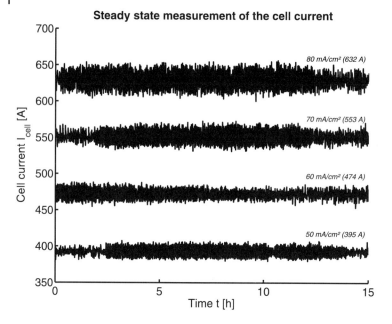

**Fig. 5.14** Measurements of the cell current at steady state for different operating points.

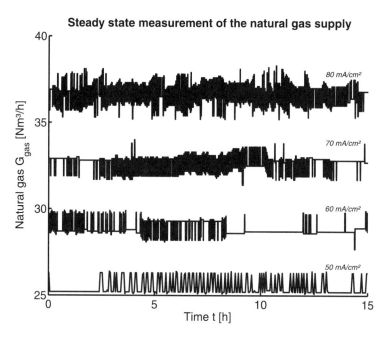

**Fig. 5.15** Measurement of the natural gas flow rate at steady state for different current densities.

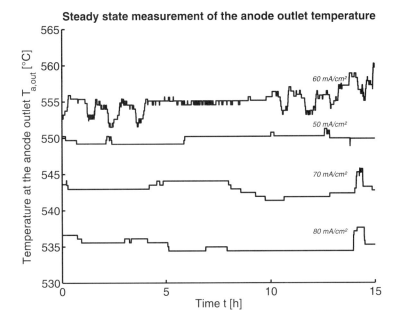

**Fig. 5.16** Measurements of the anode outlet temperature at steady state for different current densities.

are determined in this way do not only depend on the tolerance of the sensors but is mainly influenced by the effects that were explained before. Furthermore, the number and position of the sensors play an important role. All these influences were considered while estimating the measurement errors shown in Table 5.1.

### 5.3.2
### Plant Balancing and Error Minimisation

After the acquisition of the steady state measurement data all unknown states are calculated. As explained in Section 5.2.2 a balancing of the fuel cell system is not sufficient in this case. For instance, the value for the cathode reflux ratio at 50 mA/cm² differs in the range of $\Gamma_{\text{back}} = 5.7, \dots, 17.2$ depending on the chosen balance equations. Also the concentrations cannot be determined exactly like the water content at the anode inlet with $x_{H_2O, \text{iir, out}} = 0.22, \dots, 0.32$.

For a more accurate calculation of the unknown values an error minimisation is carried out as explained in Section 5.2.2. The results of this optimisation are shown in Table 5.8 for a selected operating point. The last two columns show that only minor corrections of the measured values are necessary to fulfil the mass and energy balances. For the flow rates and temperatures the deviation between corrected and measured values is even smaller than the accuracy of the measurement sensors (see Sections 5.1.4 and 5.1.2). Also the mole fractions of

**Table 5.8** Results of the minimisation of measurement errors for $I_{cell} = 50 \text{ mA/cm}^2$.

| Name | Measured value $y_h$ | Corrected value $\bar{y}_h$ | Total deviation | Relative deviation (%) |
|---|---|---|---|---|
| $\Gamma_{gas}$ | 0.1447 | 0.1445 | −0.0002 | −0.16 |
| $\Gamma_{water}$ | 0.3866 | 0.3886 | 0.002 | 0.53 |
| $\Gamma_{air}$ | 3.7646 | 3.8438 | 0.0792 | 2.1 |
| $\vartheta_{air}$ | 1.0423 | 1.0423 | 0 | 0 |
| $\vartheta_{iir, in}$ | 2.1183 | 2.118 | −0.0003 | −0.02 |
| $\vartheta_{a, out}$ | 2.7627 | 2.7629 | 0.0002 | 0.01 |
| $\vartheta_m$ | 2.8834 | 2.8864 | 0.003 | 0.1 |
| $\vartheta_{ex}$ | 2.8562 | 2.8582 | 0.002 | 0.07 |
| $x_{CH_4, iir, in}^{tr}$ | 0.5601 | 0.5674 | 0.0073 | 1.3 |
| $x_{H_2, iir, in}^{tr}$ | 0.3142 | 0.3151 | 0.0009 | 0.29 |
| $x_{CO, iir, in}^{tr}$ | 0 | 0 | 0 | 0 |
| $x_{CO_2, iir, in}^{tr}$ | 0.1091 | 0.102 | −0.0071 | −6.53 |
| $x_{N_2, iir, in}^{tr}$ | 0.0165 | 0.0155 | −0.001 | −5.91 |
| $x_{CH_4, iir, out}^{tr}$ | 0.0688 | 0.0708 | 0.002 | 2.91 |
| $x_{H_2, iir, out}^{tr}$ | 0.6766 | 0.7151 | 0.0385 | 5.69 |
| $x_{CO, iir, out}^{tr}$ | 0.0787 | 0.077 | −0.0017 | −2.18 |
| $x_{CO_2, iir, out}^{tr}$ | 0.1408 | 0.1307 | −0.0101 | −7.19 |
| $x_{N_2, iir, out}^{tr}$ | 0.0348 | 0.0065 | −0.0283 | −81.44 |
| $x_{CH_4, a, out}^{tr}$ | 0.0054 | 0.0054 | 0 | 0 |
| $x_{H_2, a, out}^{tr}$ | 0.2825 | 0.3054 | 0.0229 | 8.1 |
| $x_{CO, a, out}^{tr}$ | 0.0721 | 0.0741 | 0.002 | 2.83 |
| $x_{CO_2, a, out}^{tr}$ | 0.6287 | 0.6098 | −0.0189 | −3.0 |
| $x_{N_2, a, out}^{tr}$ | 0.0109 | 0.0052 | −0.0057 | −52.26 |
| $x_{CO_2, m}^{tr}$ | 0.0654 | 0.0662 | 0.0008 | 1.19 |
| $x_{O_2, m}^{tr}$ | 0.1576 | 0.1454 | −0.0122 | −7.72 |
| $x_{N_2, m}^{tr}$ | 0.7768 | 0.7884 | 0.0116 | 1.49 |
| $x_{CO_2, ex}^{tr}$ | 0.0462 | 0.0415 | −0.0047 | −10.1 |
| $x_{O_2, ex}^{tr}$ | 0.136 | 0.1377 | 0.0017 | 1.27 |
| $x_{N_2, ex}^{tr}$ | 0.8171 | 0.8207 | 0.0036 | 0.45 |

the fuel gas $(x_{i, iir, in}^{tr})$, the cathode inlet $(x_{i, m}^{tr})$ and the exhaust air $(x_{i, ex}^{tr})$ need only little corrections below 1 vol%, which is nearly negligible compared to the accuracy of the gas chromatograph.

In contrast, the deviations of the anode inlet mole fractions $(x_{i, iir, out}^{tr})$ are more noticeable due to the high measurement uncertainty at this point. On the one hand, there is only one measuring point in the anode gas manifold and on the other hand the leak air influences the measurements. Due to that especially the mole fractions of hydrogen $x_{H_2, iir, out}^{tr}$ and nitrogen $x_{N_2, iir, out}^{tr}$ are corrected to compensate the measurement errors. The changes of the gas compositions at the anode inlet in turn also influence a change at the anode outlet $(x_{i, a, out}^{tr})$ whereas the corrections at both measuring points are still below the expected measurement error.

The relative deviations for $x^{tr}_{N_2,\text{iir,out}}$ and $x^{tr}_{N_2,a,\text{out}}$ that seem to be very high at the first glance are also uncritical because the absolute values are very small and the measurement uncertainty is high because the leak air in the anode gas manifold mainly influences the mole fractions of nitrogen.

### 5.3.3
### Parameter Estimation

The measurements at the HotModule and the plant balancing combined with the error minimisation provide complete sets of measurement data for all operating points under investigation. The measurement data fulfill the conservation laws and will be used in the following section to estimate the model parameters.

#### Estimation of the Model Parameter $R_{\text{back}}$

The cathode recycle ratio $R_{\text{back}}$ can be determined by the molar flows within the HotModule. It is calculated by Eq. (5.6) as the ratio of the cathode recycle flow rate $\Gamma_{\text{back}}$ and the total flow rate at the cathode outlet $\Gamma_{c,\text{out}}$. The resulting parameter values are listed in Table 5.9. They range from $R_{\text{back}} = 0.62$ to $0.72$ but this result shows no feasible relation between the current density or any other input condition and the calculated recycle ratio. But one has to keep in mind that within the HotModule no flow rate measurements are directly available and the values for $\Gamma_{\text{back}}$ and $\Gamma_{c,\text{out}}$ have to be calculated by plant balancing and error minimisation. Considering these measurement uncertainties a constant value of

$$R_{\text{back}} = 0.7 \tag{5.40}$$

is chosen for the cathode recycle ratio at all operating points.

**Table 5.9** Calculated cathode recycle ratio for single operating points.

| Current density (mA/cm²) | $\Gamma_{\text{back}}$ | $\Gamma_{c,\text{out}}$ | $R_{\text{back}}$ |
|---|---|---|---|
| 50 | 9.4 | 13.8 | 0.68 |
| 60 | 12.7 | 17.7 | 0.72 |
| 70 | 13.0 | 18.5 | 0.70 |
| 80 | 10.2 | 16.5 | 0.62 |

#### Estimation of the Damköhler and Stanton Numbers

To determine the values for the most important model parameters $St_{\text{iirs}}$, $St_{\text{as}}$, $St_{\text{cs}}$, $Da_{\text{ox1}}$ and $Da_{\text{red}}$ the optimisation procedure described in Section 5.2.4 is carried out for all operating points separately in the first step. As this parameter estimation requires a number of iterations the model is in the first instance calcu-

**Table 5.10** Estimated values for the kinetic parameters at different operating points.

| Current density (mA/cm$^2$) | $St_{iirs}$ | $St_{as}$ | $St_{cs}$ | $Da_{ox1}$ | $Da_{red}$ |
|---|---|---|---|---|---|
| 50 | 38.8 | 47.4 | 155 | 6.1 | 0.141 |
| 60 | 40 | 63 | 181 | 4.3 | 0.118 |
| 70 | 38.5 | 34.5 | 126 | 8.1 | 0.113 |
| 80 | 41 | 35 | 129 | 7.0 | 0.126 |

lated with a spatial discretisation of $4 \times 4$ elements to enable faster computational times at a sufficient accuracy. The determined parameter values are afterwards used as starting values for an optimisation with a more precise discretisation of $10 \times 10$ elements. The resulting parameter values for the different operation points are listed in Table 5.10. It can be observed that the parameter values have the same order of magnitude and do correspond very well in some cases (e.g. $St_{iirs}$). Nevertheless it is not possible to determine a single set of parameters straightforward that is valid for the whole operating range.

For this purpose a further optimisation is necessary that considers the deviations between simulation an measurements for all operating points simultaneously. Therefore, the total objective function (5.34) from Section 5.2.5 has to be minimised. This is done by a consecutive calculation of single parameters and parameter groups whereas a spatial discretisation of $10 \times 10$ elements is used to enable the best possible accuracy with a reasonable computational effort.

To illustrate the parameter estimation the following figures show the total objective function $F_{total}$ depending on the particular parameters at the last optimisation step. In Fig. 5.17 the dependence on the parameter $St_{iirs}$ can be seen. The objective function has a distinct minimum at $St_{iirs} = 38$. As expected this value is in the range of the parameter values determined for the single operating points (see Table 5.10).

For the estimation of the Stanton numbers $St_{as}$ and $St_{cs}$ both one-dimensional and two-dimensional optimisations were carried out. Because in the range of the minimum the objective function is much more sensitive to changes of $St_{as}$ the results of the parameter estimation are illustrated best with the one-dimensional curves in Figs. 5.18 and 5.19. According to that the optimal parameter values for the Stanton numbers are $St_{as} = 40$ and $St_{cs} = 138$. However, the minimum here is much less pronounced than for the parameter $St_{iirs}$. Especially in the case of $St_{cs}$ the curve is very flat in the range of the optimum.

To estimate the Damköhler numbers $Da_{ox1}$ and $Da_{red}$ a two-dimensional optimisation was carried out regarding the total objective function because both parameters influence the system states in a similar manner. For an efficient numerical treatment of the problem the objective function is approximated by a polynomial. Figure 5.20 shows this approximated function as a level curve depending on the two parameters. The minimum of the objective function can be

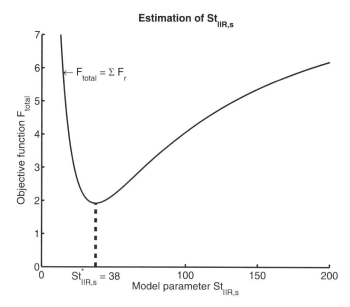

**Fig. 5.17** Dependence of the total objective function on the model parameter $St_{iirs}$. All other kinetic parameters remain constant with the values listed in Table 5.11.

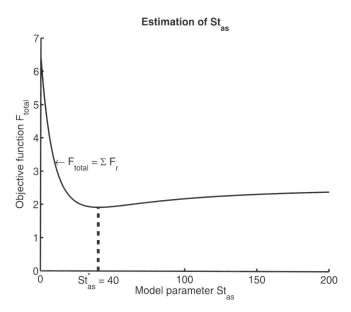

**Fig. 5.18** Dependence of the total objective function on the model parameter $St_{as}$. All other kinetic parameters remain constant with the values listed in Table 5.11.

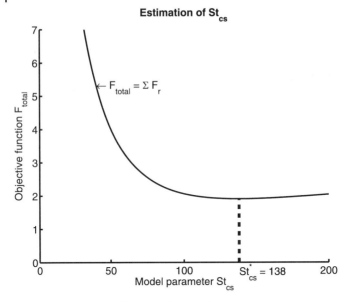

**Fig. 5.19** Dependence of the total objective function on the model parameter $St_{cs}$. All other kinetic parameters remain constant with the values listed in Table 5.11.

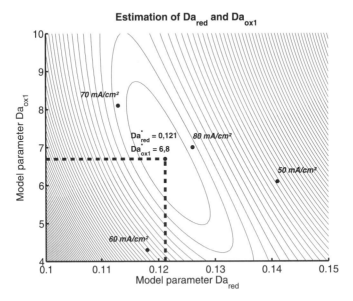

**Fig. 5.20** Total objective function as a contour plot depending on the model parameters $Da_{red}$ und $Da_{ox1}$. All other kinetic parameters remain constant with the values listed in Table 5.11.

**Table 5.11** Estimated values of the kinetic parameters for the whole operating range.

| Current density | $St_{iirs}$ | $St_{as}$ | $St_{cs}$ | $Da_{ox1}$ | $Da_{red}$ |
|---|---|---|---|---|---|
| 50–80 mA/cm$^2$ | 38 | 40 | 138 | 6.8 | 0.121 |

found at $Da_{red} = 0.121$ and $Da_{ox1} = 6.8$. For comparison the optimal parameter configurations at the single operating points that are listed in Table 5.10 are also marked in the diagram. As expected the total minimum is located between these points. The diagram also shows that the gradient of the objective function for $Da_{red}$ is higher than for $Da_{ox1}$. This corresponds to the results of the sensitivity analysis in Section 5.2.3.

All kinetic parameters that were estimated by the previous optimisations are outlined in Table 5.11.

### 5.3.4
### Dynamic Measurements

For the analysis of the dynamic behaviour of the HotModule two measurements have been carried out at load changes, i.e. at changing cell currents. One measurement was done at 60 mA/cm$^2$ → 70 mA/cm$^2$ and the other at 70 mA/cm$^2$ → 80 mA/cm$^2$.

The cell current does not increase instantaneously at the HotModule but in several small steps within 10–15 min. Figure 5.21 shows the cell current during the load change from $I_{cell} = 70$ mA/cm$^2$ ($\cong 553$ A) to $I_{cell} = 80$ mA/cm$^2$ ($\cong 632$ A). As already observed at the steady states the cell current also oscillates in dynamic operation with up to $\pm 15$ A regarding the setpoint. This is due to external disturbances and the influence of the plant control.

Simultaneously with the increasing cell current also the fuel gas feed is adjusted. This is also done in small steps whereas the fuel gas is still controlled after the load change as can be seen in Fig. 5.22.

Due to the oscillations of the input variables also the cell voltage in Fig. 5.23 shows a significant noise. On the basis of this measurement it is not clearly observable at what time a new steady state is reached. In this regard the temperature measurements in Fig. 5.24 provide a better information. Besides the temperatures at the anode and cathode outlet the diagram also shows a temperature measurement $T_{Stack}$ within the cell stack. The absolute value is not very significant because it is only a single measurement within a non-representative cell but the curves show a good tendency at what time a new steady state is reached. This is the case at approximately $t = 4, \ldots, 5$ h, thus 3 to 4 h after the load change. The subsequent fluctuations are caused by external disturbances as already observed for the steady state measurements.

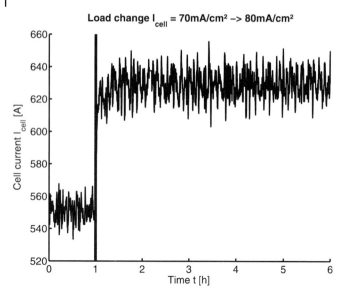

**Fig. 5.21** Measurement of the cell current during the load change from $I_{cell} = 70$ mA/cm$^2$ ($\cong 553$ A) to $I_{cell} = 80$ mA/cm$^2$ ($\cong 632$ A) at $t = 1$ h.

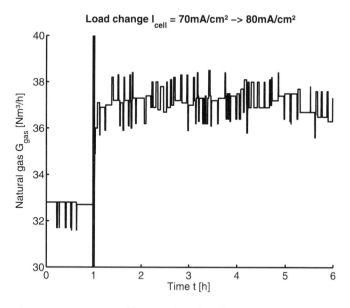

**Fig. 5.22** Measurement of the natural gas feed flow rate during the load change from $I_{cell} = 70$ mA/cm$^2$ ($\cong 553$ A) to $I_{cell} = 80$ mA/cm$^2$ ($\cong 632$ A) at $t = 1$ h.

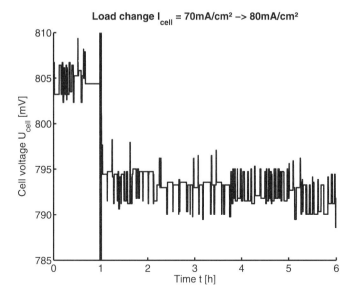

**Fig. 5.23** Measurement of the cell voltage during the load change from $I_{cell} = 70$ mA/cm$^2$ ($\cong 553$ A) to $I_{cell} = 80$ mA/cm$^2$ ($\cong 632$ A) at $t = 1$ h.

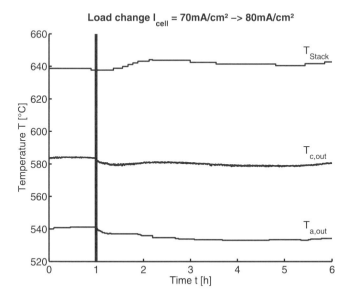

**Fig. 5.24** Measurement of selected temperatures during the load change from $I_{cell} = 70$ mA/cm$^2$ ($\cong 553$ A) to $I_{cell} = 80$ mA/cm$^2$ ($\cong 632$ A) at $t = 1$ h.

### 5.3.5
### Estimation of the Solid Heat Capacity

The different physical and chemical phenomena that occur in an MCFC are governed by different time constants. In the model equations they are represented by the particular capacities for charge, mass and heat energy. Of all processes in the MCFC the temperature change of the solid phase is by far the slowest one. This is very obvious as the HotModule can be seen as a block of steel with several tons of weight. If this is heated up or cooled down it takes quite long. This means that the temperature dynamics dominates all other processes because the relevant system states like cell voltage, gas temperatures and gas concentrations depend directly or indirectly on $T_s$. The model parameter that influences the time constant of the temperature dynamics is the heat capacity $c_{p,s}$ of the solid phase. It has to be estimated with the help of the measurements discussed in Section 5.3.4.

The heat capacity has to be chosen in such a way that the time that the system needs to reach a new steady state in the model simulations is approximately 3 to 4 h as determined from the measurements. For this purpose model simulations with different values of $c_{p,s}$ have been carried out. Figure 5.25 shows the temperature at the anode outlet during a simulated load change for different values of the model parameter $c_{p,s}$. Thereby it is important to take into consideration that simulated and experimental load change differ in one important aspect. Whereas

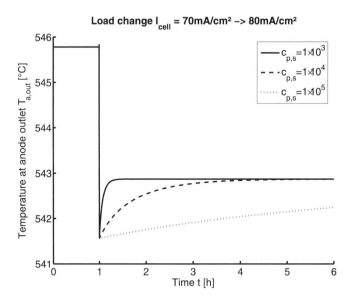

**Fig. 5.25** Simulated anode outlet temperature during a load change from $I_{cell} = 70$ mA/cm$^2$ to $I_{cell} = 80$ mA/cm$^2$ at $t = 1$ h for different values of the heat capacity $c_{p,s}$.

the load change at the HotModule is carried out in small steps within several minutes, in the simulation the input conditions are changed promptly to the new setpoint at $t = 1$ h. Therefore, the simulated temperature in Fig. 5.25 and the measured curves in Fig. 5.24 show qualitative differences. In particular the undershooting behaviour shortly after the load change that can be observed in the simulations is not visible in the measurements. But for a more realistic simulation of the load change the gas preparation unit would have to be included in the MCFC model which was not part of this investigation.

Apart from that the influence of the heat capacity on the temperature dynamics in Fig. 5.25 can be clearly seen. The time constant of this process strongly depends on this parameter. The simulations show that with a value of $c_{p,s} \approx 10,000$ the time until the new steady state is reached is 3 to 4 h which corresponds best to the experiments.

### 5.3.6
### Evaluation of the Results

**Analysis of the Damköhler Numbers of the Reforming Reactions and the CO Oxidation**

In Section 5.2.3 it has been assumed that the reforming reactions within the IIR and the anode channels are close to equilibrium. Due to that the corresponding dimensionless reaction rate constants were allocated sufficiently high values:

$$Da_{iir, ref1} = Da_{iir, ref2} = Da_{a, ref1} = Da_{a, ref2} = 200 \qquad (5.41)$$

In the case of the IIR the assumption of equilibrium can be verified with the concentration and temperature measurements at the anode inlet. The reaction kinetics of the methane steam reforming (ref1) and the water–gas shift reaction (ref2) are described in the MCFC model [5] as follows:

$$r_{iir, ref1} = k_{iir, ref1} \underbrace{\left( \chi_{CH_4, iir} \cdot \chi_{H_2O, iir} - \frac{\chi_{CO, iir} \cdot \chi_{H_2, iir}^3}{K_{ref1}(\vartheta_{iir})} \right)}_{X_{ref1}} \qquad (5.42)$$

$$r_{iir, ref2} = k_{iir, ref2} \underbrace{\left( \chi_{CO, iir} \cdot \chi_{H_2O, iir} - \frac{\chi_{CO_2, iir} \cdot \chi_{H_2, iir}}{K_{ref2}(\vartheta_{iir})} \right)}_{X_{ref2}} \qquad (5.43)$$

Thereby for the temperature dependence of the reaction rate an exponential approach according to Arrhenius is chosen:

$$k_j = \exp\left[ Arr_j \left( \frac{1}{\vartheta_j^0} - \frac{1}{\vartheta_{iir}} \right) \right] \qquad (5.44)$$

**Table 5.12** Reaction rate constants of the reforming reactions at the outlet of the indirect internal reformer. Negative values of $X_j$ indicate that the equilibrium has been exceeded.

| Current density (mA/cm²) | $K_{ref1}$ | $X_{ref1}$ | $K_{ref2}$ | $X_{ref2}$ |
|---|---|---|---|---|
| 50 | 0.1589 | −0.0358 | 3.2933 | $0.1546 \times 10^{-3}$ |
| 60 | 0.1423 | −0.0392 | 3.3534 | $-0.6914 \times 10^{-3}$ |
| 70 | 0.1391 | −0.0367 | 3.3657 | $1.5993 \times 10^{-3}$ |
| 80 | 0.1241 | −0.0345 | 3.4291 | $-0.4330 \times 10^{-3}$ |

As this expression is always positive the direction of the reactions is determined by $X_{ref1}$ and $X_{ref2}$. These expressions account for the dependence of the reaction rate on the concentrations. They contain the equilibrium constants $K_{ref1}$ and $K_{ref2}$ that are defined as

$$K_j(\vartheta_{iir}) = \exp\left(\Delta_R s_j^0(\vartheta_{iir}) - \frac{\Delta_R h_j^0(\vartheta_{iir})}{\vartheta_{iir}}\right) \tag{5.45}$$

By putting the measurements of $x_{i,iir,out}$ and $\vartheta_{iir,out}$ into the reaction kinetics the appropriate values of $K_j$ and $X_j$ can be obtained for different operating points. These values are listed in Table 5.12. The results show that for all operating points the methane steam reforming is already beyond equilibrium ($X_{ref1} < 0$) and in two cases the water–gas shift reaction seems to run backwards as well ($X_{ref2} < 0$). The main reason for the negative reaction rates is the input of the leak air into the anode gas manifold as explained in Section 5.2.3.

All in all this analysis shows particularly that the reforming reaction within the IIR and the anode channels run very fast and they are close to equilibrium.

Furthermore, using model simulations it can be demonstrated that the parameter values in Eq. (5.41) make sense. Figure 5.26 shows the objective function (Eq. (5.34)) depending on the particular Damköhler numbers whereas for all other parameters the values determined in Section 5.3.3 are used. The diagrams show that smaller Damköhler numbers do not result in a lower value of the objective function that means no better correlation between simualtion and measurement can be achieved.

In Section 5.2.3 it has also been discussed that the oxidation of carbon monoxide at the anode (ox2) can be neglected for the investigated operating range. Therefore the corresponding Damköhler number was set to $Da_{ox2} = 0$. Also in this case the variation of this parameter shows that the consideration of the CO oxidation with $Da_{ox2} > 0$ does not bring any improvement of the objective function as can be seen in Fig. 5.27.

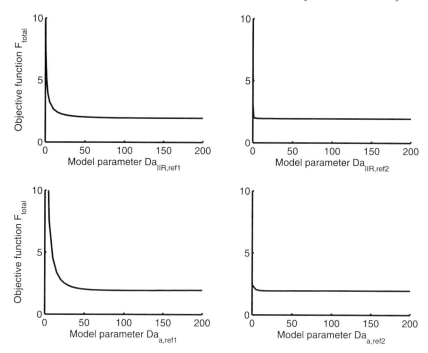

**Fig. 5.26** Dependence of the total objective function on single Damköhler numbers of the reforming reactions.

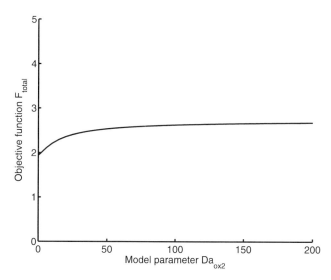

**Fig. 5.27** Dependence of the total objective function on the Damköhler number $Da_{ox2}$ of the carbon monoxide oxidation.

**Table 5.13** Deviation between model simulation and measurements for different operating points compared to the measurement accuracy.

|  | 50 mA/cm$^2$ | 60 mA/cm$^2$ | 70 mA/cm$^2$ | 80 mA/cm$^2$ | Measurement accuracy |
|---|---|---|---|---|---|
| $\Delta \tilde{U}_{\text{cell}}$ | −6 mV | +7 mV | +3 mV | −4 mV | ±5 mV |
| $\Delta T_{\text{iir, out}}$ | +3.3 K | −0.8 K | −2.5 K | −4.0 K | ±30 K |
| $\Delta T_{a,\text{out}}$ | −1.0 K | −13 K | +2.1 K | +6.9 K | ±20 K |
| $\Delta T_m$ | +0.2 K | −12 K | −5.0 K | −5.6 K | ±10 K |
| $\Delta T_{\text{ex}}$ | +2.9 K | −5.1 K | −1.3 K | +1.8 K | ±3 K |
| $\Delta x_{i,\text{iir, out}}$ | <5% | <6% | <6% | <6% | ±5% |
| $\Delta x_{i,a,\text{out}}$ | <1% | <1% | <2% | <1% | ±5% |
| $\Delta x_{i,m}$ | <2% | <2% | <3% | <4% | ±5% |
| $\Delta x_{i,\text{ex}}$ | <1% | <1% | <1% | <1% | ±2% |
| $F_r$ | 0.70 | 0.80 | 0.14 | 0.27 | |

**Comparison of Simulations and Measurements at Steady State**

In the previous chapters the values for all parameters have been determined that influence the steady state behaviour of the MCFC. Now it is to be reviewed how good the model simulations agree with the actual measurements. Therefore the cell voltage, the most important temperatures and the concentration measurements of the four investigated operating points are compared to the simulation results.

Table 5.13 shows the differences between simulations and measurements for all relevant measurement data. For all operating points the deviations are in the same order of magnitude whereas the best agreement is achieved at 70 mA/cm$^2$. Here the total objective function $F_r$ shows the lowest value. The largest deviation can be observed for 60 mA/cm$^2$, which is mainly due to the difference of the cell voltage $\Delta \tilde{U}_{\text{cell}}$ that has the strongest influence on the objective function (see Section 5.2.4). When analysing Table 5.13 it is very important to compare the deviations to the particular measurement accuracy of the sensors. The differences between simulation and measurements have the same order of magnitude as the measurement accuracy and for some values they are even well below. This leads to the conclusion that the measurements are very well reflected by the model simulations.

Besides the measurements at 50, 60, 70 and 80 mA/cm$^2$ a fifth measurement was done at $I_{\text{cell}} = 100$ mA/cm$^2$ after the completion of the parameter estimation. This gives the possibility to check whether the model is valid beyond the investigated operating range.

But the measurements at 100 mA/cm$^2$ differ in two aspects from the other operating points. First, this current density is at the upper limit that is possible at the HotModule in Magdeburg due to technical restrictions for instance by the

**Table 5.14** Deviation between model simulation and measurements compared to the measurement accuracy for an operating point that was not included in the parameter estimation.

| | 100 mA/cm² | Measurement accuracy |
|---|---|---|
| $\Delta \tilde{U}_{cell}$ | +6 mV | ±5 mV |
| $\Delta T_{iir,out}$ | −6.9 K | ±30 K |
| $\Delta T_{a,out}$ | +16.7 K | ±20 K |
| $\Delta T_m$ | +9.9 K | ±10 K |
| $\Delta T_{ex}$ | −2.7 K | ±3 K |
| $\Delta x_{i,iir,out}$ | <9% | ±5% |
| $\Delta x_{i,a,out}$ | <4% | ±5% |
| $\Delta x_{i,m}$ | <3% | ±5% |
| $\Delta x_{i,ex}$ | − | ±2% |
| $F_r$ | 0.59 | |

electrical inverters. Therefore, the fuel cell only stayed in the operating mode for a rather short period of time and the measurements had to start 3 h after the load change. At this point the steady state had not been fully reached. Secondly the mole fractions at the cathode outlet ($x_{ex}$) could not be measured due to a plugged intake pipe. So these values are not available for balancing and comparison with the simulations.

Table 5.14 shows the deviations between simulations and measurements for $I_{cell} = 100$ mA/cm². Also in this case they are in the same range as at the other operating points. Especially the cell voltage which is the most important measurement shows a good agreement. But for the concentrations the difference between simulation and measurement is highest for this operating point whereas an additional measurement uncertainty has to be considered due to the missing measurement of $x_{ex}$. Because the concentrations are weighted fairly low in the target function $F_r$ its value for $I_{cell} = 100$ mA/cm² is even below 50 mA/cm² and 60 mA/cm².

Despite the more complicated measurement conditions for this load case it can be found that the model shows a good agreement with the measurement data also for $I_{cell} = 100$ mA/cm². As this operating point is clearly above the range for which the parameter estimation was done it can be assumed that the model can be extrapolated quite good.

To illustrate the small deviations between the model simulations and the experimental data, Fig. 5.28 shows the cell voltages for all operating points as a current–voltage curve. This again points out the good agreement of this important state over a wide operating range.

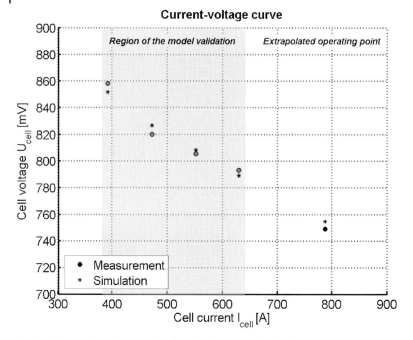

**Fig. 5.28** Comparison between simulated and measured cell voltages for different operating conditions.

## 5.4
## Summary

In this section the identification of the parameters from the MCFC model was presented. This process required an elaborate strategy for several reasons. First, although the plant is equipped with a large array of sensors (mass flow meters, thermocouples and a gas chromatograph), measurement information is incomplete and not always as reliable as it could be under laboratory conditions. Secondly, with more than 30 unknown parameters, the number of fitting parameters is rather high. Thirdly, the parameters must be fitted so that not only a single operating point, but a whole range of load cases is correctly reflected by the model.

To handle this challenge, a four-step strategy is applied. First, a black box modelling of several compartments of the system was used to complete and correct the measurement data obtained from several load cases covering a whole operating range. In this step, compliance of the measurement data with the conservation laws of mass and energy was established and the reliability and uncertainty of each single measurement were taken into account. The second aspect was a parameter sensitivity analysis that helped to eliminate insensitive parameters which could be set to arbitrary high or low values, depending on the parameter. Thus five relevant parameters could be identified that mainly influence the steady state behaviour and one more parameter determining the temperature dynamics.

In the third step the iterative adaption of parameter groups was carried out for each set of measurement data separately. This led to several individual parameter sets. Finally, the simultaneous adaption of the parameters for all data sets results in one single parameter set for all measured load cases. This could be achieved by using an approximated error function.

The simulations of the validated MCFC model show a very good agreement with the measurements from the HotModule. The deviations are in the range of the measurement accuracies, or even below. Furthermore, the model shows good extrapolation properties.

## Bibliography

1 Datenblatt Mantel-Thermoelemente, SAB BRÖCKSKES GmbH & Co. KG, Grefrather Str. 204-212b, D-41749 Viersen, http://www.sab-broeckskes.de/

2 Datenblatt Proline promass 80F, Endress+Hauser Messtechnik GmbH+Co. KG, Colmarer Strasse 6, D-79576 Weil am Rhein, http://www.de.endress.com/

3 Datenblatt Durchflussmesser PEL, KOBOLD Messring GmbH, Nordring 22-24, D-65719 Hofheim/Ts., http://www.koboldmessring.com/

4 Datenblatt Strömungssensor SS 20.60, SCHMIDT Technology GmbH, Feldbergstrasse 1, D-78112 St. Georgen, http://www.schmidt-feintechnik.de/

5 P. Heidebrecht, Modelling, Analysis and Optimisation of a Molten Carbonate Fuel Cell with Direct Internal Reforming (DIR-MCFC), Fortschritt-Berichte, VDI-Verlag, Düsseldorf, 2005.

6 C.Y. Yuh, J.R. Selman, The polarization of molten carbonate fuel cell electrodes: 1. Analysis of steady-state polarization data, Journal of The Electrochemical Society 138, 1991, 3642–3648.

7 J. Xu, G.F. Froment, Methane-steam reforming, methanation and water gas shift: I. Intrinsic kinetics, AIChE Journal 35, 1989, 88–96.

8 K. Hou, R. Hughes, The kinetics of methane steam reforming over a Ni/a-Al2O catalyst, Chemical Engineering Journal 82, 2001, 311–328.

# 6
# Steady State and Dynamic Process Analysis

*Peter Heidebrecht, Matthias Gundermann, and Kai Sundmacher*

In this chapter, the numerical solution of the two-dimensional reference model (see Chapter 3) is presented. In the first part, selected steady state simulation results at one chosen operating point are shown and discussed. These results also include a current–voltage curve for an isothermal cell model. Afterwards, a virtual load change is simulated. Simulation results from a model without the indirect internal reformer can be found in [1, 2].

## 6.1
## Steady State Simulation

To solve the reference model its PDEs (partial differential equations) are first transformed into a system of algebraic and ordinary differential equations (DAE) by finite volume discretisation with respect to both spatial coordinates. The convective terms are discretised using the upwind scheme with respect to the particular flow direction in each compartment. Special attention is required with the discretisation of the temperature equations. The discretised form must be conservative with respect to the law of energy conservation must be fulfilled. The resulting equations are implemented in ProMoT (Process Modelling Tool, version 0.7.0, [3]) and solved in the simulation environment DIVA (version 3.9, [4]).

The operating conditions imposed on the reference model in this simulation correspond to those of the 'HotModule' at 80 mA/cm$^2$ and are listed in Table 6.1. The model parameters used are given in Chapter 5. All plots are oriented in the same way as indicated in Fig. 6.1. The results are obtained using a $10 \times 10$ spatial discretisation.

Figures 6.1 and 6.2 show two selected states within the indirect internal reforming compartment. A slightly pre-reformed gas mixture (about 15% hydrogen) enters this compartment at low temperature (the dimensionless inlet temperature of 2.2 corresponds to 656 K). As both figures show, the gas is heated by the surrounding solid cell compartments and is further reformed while flowing through the reformer channels. The gas at the reformer outlet is close to the

*Molten Carbonate Fuel Cells.* Edited by Kai Sundmacher, Achim Kienle, Hans Josef Pesch, Joachim F. Berndt, and Gerhard Huppmann
Copyright © 2007 WILEY-VCH Verlag GmbH & Co. KGaA, Weinheim
ISBN: 978-3-527-31474-4

**Table 6.1** Steady state operating conditions used in this section.

| Feed gas | | | | | | | |
|---|---|---|---|---|---|---|---|
| $\Gamma_{iir, in}$ | 0.83 | | | | | | |
| $\vartheta_{iir, in}$ | 2.17 | | | | | | |
| $\chi_{i, iir, in}$ | $CH_4$ | $H_4O$ | $H_2$ | CO | $CO_2$ | $O_2$ | $N_2$ |
| | 0.222 | 0.574 | 0.151 | 0 | 0.047 | 0 | 0.006 |
| Air | | | | | | | |
| $\Gamma_{air}$ | 5.49 | | | | | | |
| $\vartheta_{iir, in}$ | 1.08 | | | | | | |
| Cathode recycle | | | | | | | |
| $R_{back}$ | 0.7 | | | | | | |
| Cell current | | | | | | | |
| $I_{cell}$ | 0.4473 | | | | | | |

chemical equilibrium of the reforming reactions and contains about 42 mole% of hydrogen. Both diagrams indicate that all channels within the reformer compartment show nearly identical behaviour.

Figures 6.3 to 6.5 show some states from the anode gas phase. The mole fraction of hydrogen is plotted in Fig. 6.3. In the anode channel, hydrogen is consumed by the electrochemical oxidation. Simultaneously, new hydrogen is produced by the reforming process. Due to the extensive reforming reaction in the external and indirect internal reforming compartments, the reforming process is at equilibrium at the inlet of the anode channels. The activity of the reforming catalyst is high enough to keep the anode gas close to the equilibrium throughout the whole channel. The molar fraction of hydrogen at the outlet is at about 12%,

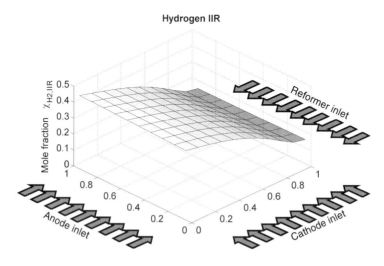

**Fig. 6.1** Molar fraction of hydrogen in the indirect internal reforming channels.

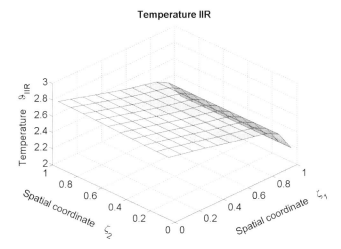

**Fig. 6.2** Dimensionless gas temperature in the indirect internal reforming channels.

and the methane concentration (not shown here) is at about 1%. The hydrogen profile is not identical in all channels; the channels on the left-hand side ($\zeta_2 = 1$) have a higher hydrogen consumption than the rest of the channels. This shows the distinct two-dimensional character of the HotModule.

In Fig. 6.4, the anode gas temperature is almost constant. This temperature is mainly governed by two phenomena: The cooling effect of the endothermic reforming reaction and the heat exchange with the solid parts of the cell, whose temperature is depicted in Fig. 6.8. A comparison between Figs. 6.4 and 6.8

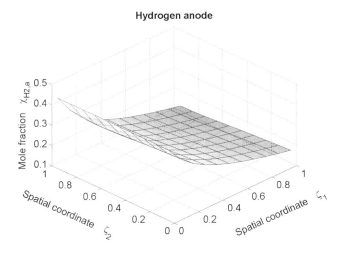

**Fig. 6.3** Molar fraction of hydrogen in the anode channels.

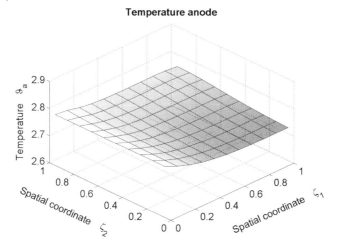

**Fig. 6.4** Dimensionless gas temperature in the anode channels.

shows that in the first half of the anode channels ($\zeta_1 = 0, \ldots, 0.5$), the anode gas is heated up by the solid parts. The reforming process compensates this temperature increasing effect, so that the gas temperature remains nearly constant. Closer to the channel outlet, the anode and the solid temperature are almost equal, so the heat exchange rate between both is rather low. Concerning the reforming process in this region, methane is almost depleted there, so the reforming process also slows down considerably. Thus the anode gas temperature is almost constant also in this region. Finally, Fig. 6.5 shows the profile of the molar

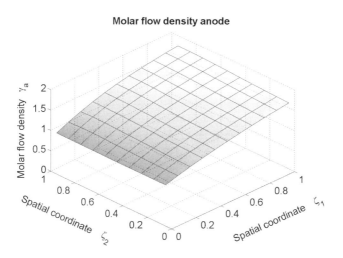

**Fig. 6.5** Dimensionless molar flow density in the anode gas channels.

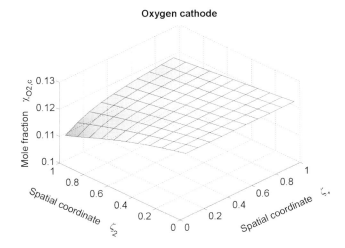

**Fig. 6.6** Molar fraction of oxygen in the cathode channels.

flow density of the anode gas. Due to the positive change in mole numbers in the reforming process and the oxidation reactions, the molar amount of gas is increased while flowing through the anode channels, which results in a significantly increased molar flow rate at the anode outlet region compared to the anode inlet.

Two states from the cathode gas phase are plotted in Figs. 6.6 and 6.7. Gas with about 12% oxygen concentration enters the cathode channels from the catalytic burner at $\zeta_2 = 0$. While flowing along the cathode channels, this concentration

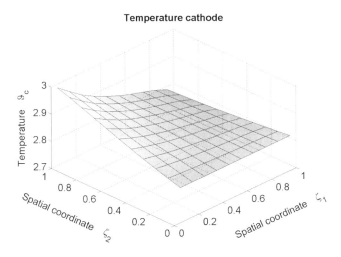

**Fig. 6.7** Dimensionless gas temperature in the cathode channels.

**Temperature solid**

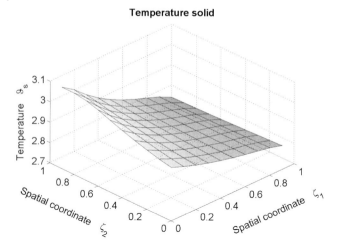

**Fig. 6.8** Dimensionless temperature of the solid parts of the cell.

is decreased due to the electrochemical reduction reaction. Because of an inhomogeneous reaction rate, the consumption of oxygen on the left-hand side channels (at $\zeta_1 = 0$) is higher than that on the right-hand side channels ($\zeta_1 = 1$). At the outlet of the cathode channels, the molar fraction of oxygen is decreased by about 1%. The molar fraction of carbon dioxide, which is the second educt in the reduction reaction, looks similar. There, the molar fraction starts at about 6% and ends at about 2% below that. Figure 6.7 shows the temperature profile of the cathode gas. Starting at a dimensionless inlet temperature of about 2.8 (560 °C), the gas temperature is only governed by heat exchange with the solid cell parts. This leads to a steadily increasing gas temperature along the channels on the left-hand side up to a maximum dimensionless temperature of about 3 (620 °C). Here the cathode gas works as a cooling agent for the solid parts. Only in the channels on the right-hand side, the cathode gas temperature is slightly decreasing. A comparison with Fig. 6.8 shows that the solid is only slightly colder than the cathode gas, so the cathode gas is cooled here.

Probably the most interesting state with respect to favourable cell operation is the temperature in the solid cell parts, which is shown in Fig. 6.8. It shows a distinct minimum at $\zeta_1 = 1$ and a strong maximum at the anode inlet/cathode outlet corner ($\zeta_1 = 0$, $\zeta_2 = 1$). This temperature profile is governed by several phenomena of heat sources and heat sinks as well as convective energy transport. The electrochemical reactions take place all over the whole cell area and act as a heat source in the solid parts of the cell. Heat sinks can be identified at the inlet region of the indirect internal reformer and in the first half of the anode channel. In the indirect internal reformer, the gas enters the cell at comparably low temperature. Upon entering the cell, it is heated up, draining heat from the solid cell parts. As its temperature increases, a considerable amount of the reforming process takes place near the reformer inlet, which keeps the gas temperature low and maintains a significant heat flow from the solid cell parts. Reforming also occurs in the anode channels. But because the gas is almost at chemical equilibrium at the anode inlet, the reaction rate is rather low compared to its rate at the inlet of the indirect internal reformer. In the anode channel, the reforming reaction mostly follows the chemical equilibrium, which is continuously perturbed by the consumption of hydrogen by the oxidation reaction. Thus a cooling effect exists in the anode channel, but this effect is neither strong nor it is focused on a special region of the cell.

Convective energy transport occurs along the flow directions of each gas phase. The heat transport capacity in the indirect internal reformer and in the anode is relatively low because of the small molar flow densities. The molar flow in the cathode channels is considerably higher and also the heat exchange coefficient between the cathode gas and the solid parts, $St_{cs}$, is three times higher than that in the anode. Thus the convective energy transport in the cathode channel is the dominant transport mechanism, moving energy along $\zeta_2$. Heat conduction in the solid cell parts is also present, but it is not a directed way of energy transport. Its effect is simply that it tends to balance out temperature gradients. In the present model, the heat conductivity of the solid parts is relatively low, so that heat conduction only plays a minor role.

With these considerations in mind, the cell has a strong cooling effect due to the indirect internal reforming at $\zeta_1 = 0.5, \ldots, 1$, and a dominant convective heat transport along $\zeta_2$ due to the cathode gas flow. This leads to a temperature minimum at $\zeta_1 = 1$ and a distinct maximum temperature at $\zeta_1 = 0$, $\zeta_2 = 1$. The dimensionless temperature difference between the hot spot and the cold region is at about 0.3, which corresponds to 90 K.

The local rates of the electrochemical reactions determine the electric current density. They depend on the local gas concentrations as well as on the cell tem-

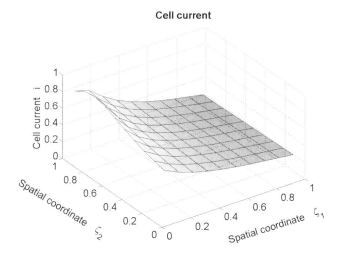

**Fig. 6.9** Dimensionless cell current density.

perature. Figure 6.9 shows a clear maximum of the electric current density at the hot spot despite the fact that anode hydrogen concentrations are low in this regions. This indicates that the influence of the solid temperature is dominant over the effect of concentration changes, at least at concentration levels which are typical for MCFC.

## 6.2
## Current–Voltage Curve

Like the spatially distributed profiles discussed in the previous section, the current–voltage curve is an important steady state characteristic of a fuel cell. In principle, current–voltage curves can be produced under different conditions.

- The operational current–voltage curve comprises points from operating conditions of a real fuel cell. Not only the cell current, but also input parameters such as the amount of fuel and air are changed from one set to another. Usually only a few of these load cases are defined and they do not cover very high and very low cell currents. Thus this curve consists of only a few points in the medium range of the current–voltage curve. It is relevant for the operation of a fuel cell system. Because a whole parameter set is changed instead of only one single parameter, the physical interpretation of this curve is very difficult, if not impossible.
- A different current–voltage curve is obtained by changing the cell current, while keeping all other input parameters constant. At very low and very high currents, this leads to very low efficiencies. In real fuel cell systems, this comes with a high amount of heat production, thus high temperatures, and with this the destruction of the cell. To avoid the loss of the cell, the experimental investigation of this curve is limited to a very narrow region of cell currents. The simulation of the cell behaviour under high temperature conditions is possible, but not realistic. Thus this curve is incomplete, but it can be evaluated from a real system and it is interpretable, because only a single parameter is changed.
- The full current–voltage curve can be measured at constant cell temperature, which can only be realised in well-tempered laboratory experiments, but not with full systems. Here, all input conditions are kept constant and only the cell current is changed over its full range. The result does not necessarily resemble the operational curve, but nevertheless it is suitable for the analysis of the cell performance. The simulation of this curve requires an isothermal cell model, where the solid parts of the cell have a well-defined, constant temperature.

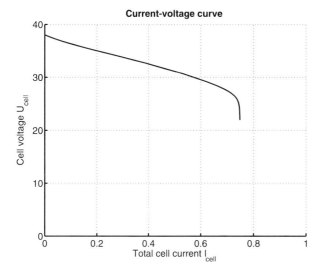

**Fig. 6.10** Dimensionless current–voltage curve obtained from an isothermal cell model.

As an example, the cell model has been applied to calculate an isothermal current–voltage curve. The input parameters have been set according to the operating conditions at a cell current of 80 mA/cm$^2$ (dimensionless $i_{cell} = 0.45$, see Table 6.1). The cell temperature was set to a constant level of 621 °C (dimensionless $\vartheta = 3$) and the cell current was varied from zero until the cell voltage started to decrease steeply.

The resulting current–voltage curve is shown in Fig. 6.10. The typically logarithmic activation region at very low cell current is not very pronounced. This is possibly due to two reasons. First, this is a spatially distributed system, and due to the concentration gradients the electrochemical reactions are not activated simultaneously at all locations in the cell. Secondly, the cell voltage is not only governed by the electrochemical process alone, but it is also influenced by the reforming reaction. This leads to an almost linear curve in this region. The limiting current corresponds to the stoichiometric limiting current, at which virtually all fuel is reformed and electrochemically consumed.

## 6.3
## Transient Simulation

The MCFC model is not only capable to predict steady states, but it is also suitable for transient simulations. As an example, a stepwise load change is demonstrated here. Using the steady state described in the previous section as an initial condition, the average electric cell current demand is suddenly increased from $I_{cell} = 0.447$ (corresponding to 80 mA/cm$^2$) to $I_{cell} = 0.560$ (corresponding to 100 mA/cm$^2$) at the dimensionless time $\tau = 0.1$. All other input parameters

**Current density**

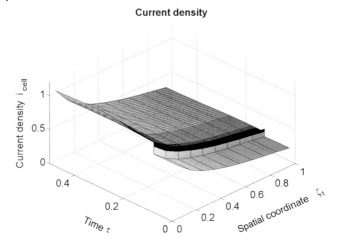

**Fig. 6.11** Short time behaviour of the electric current density after a stepwise load change at $\tau = 0.1$. The states along a middle anode channel are displayed.

such as amount of fuel gas and air are kept constant throughout the whole simulation. This allows for a reasonable physical interpretation of the simulation results. Realistically, a load change would not be imposed as a stepwise change, but the cell would rather be increased according to a ramped profile. Also, all other input parameters would be changed accordingly. Thus, the effects of all these changes would overlap and thereby render any interpretation of the cell behaviour impossible.

Figure 6.11 shows the electric current density distribution along a middle anode channel during the load change. When the current demand is increased (at dimensionless time $\tau = 0.1$), the current density increases over the whole cell area virtually instantaneously. During the next few seconds, the current density further changes slightly due to concentration changes in the gas channels. At the end time of this plot, which is at $\tau = 0.5$ (corresponding to about 5 s real time after the load change), the current density reaches a quasi steady state. In fact, the current density continues to change after this time due to processes in the cell with considerably larger time constants.

Figure 6.12 shows the hydrogen concentration in the anode channel. Due to the higher current demand, the hydrogen consumption is increased. Because methane concentrations are already low at the initial state, the reforming process cannot compensate this effect. The molar fraction of hydrogen decreases smoothly until it reaches a quasi steady state after a few seconds, which correspond to the mean residence of the gas in the anode channels.

The molar flow density of the anode gas shows a different behaviour (Fig. 6.13). The stepwise load change comes with a stepwise increase of the electrochemical oxidation reactions. Because these reactions increase the amount of gas mole-

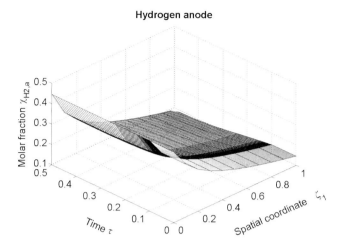

**Fig. 6.12** Short time behaviour of the mole fraction of anodic hydrogen after a stepwise load change at $\tau = 0.1$. The states along a middle anode channel are displayed.

cules, the molar flow density is also instantaneously increased along the whole anode channel. After a short time, it reaches a new quasi steady state profile.

Directly after the load change, the anode exhaust gas is fed into the catalytic combustion chamber at a higher flow rate. This leads to an increased oxygen consumption from the combustion air and thus to a decreased mole fraction of oxy-

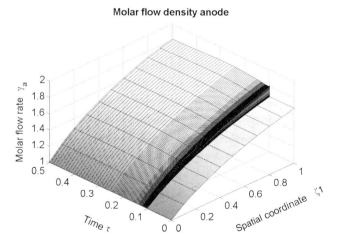

**Fig. 6.13** Short time behaviour of the anodic molar flow density in the anode after a stepwise load change at $\tau = 0.1$. The states along a middle anode channel are displayed.

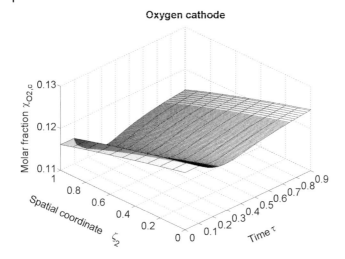

**Fig. 6.14** Short time behaviour of the mole fraction of the cathodic oxygen after a stepwise load change at $\tau = 0.1$. The states along a middle cathode channel are displayed.

gen in the cathode channel. This can be seen in Fig. 6.14 at the cathode inlet ($\zeta_2 = 0$). After a certain time, the anode outlet gas is oxidised to a higher degree by the electrochemical oxidation reactions. At that time, less oxygen is consumed in the combustion and the oxygen inlet concentration in the cathode channels increases again. At the dimensionless time $\tau = 0.9$ (which is approximately 10 s after the load change, corresponding to the mean residence time of the anode channels, the reversal chamber and the cathode channels), the mole fraction reaches a quasi steady state. Because at this time the oxygen consumption in the combustion chamber is lower than that under initial conditions, the oxygen inlet concentration is even higher than before the load change. Along the cathode channels, the molar fraction of oxygen is decreased. Because the load change does not affect the total mass balances of chemical species around the whole system, the oxygen outlet concentration at the new steady state must be identical to the outlet concentration under initial conditions.

The dynamics of the anode outlet also influence the cathode gas temperature. When the anode molar outlet flow is increased, a larger amount of combustibles is swept into the combustion chamber, while the flow air is kept constant. This leads to a higher combustion temperature, so the cathode inlet temperature is increased (see Fig. 6.15). After some time, the anode outlet gas is oxidised to a higher degree by the electrochemical oxidation reactions, so the combustion temperature and thereby the cathode inlet temperature decrease again. This is an inverse system behaviour which complicates the prediction of system behaviour and the development of process control systems. Furthermore, the dynamics of both cathode states shown document the importance of the coupling between the anode and the cathode channels via the combustion chamber.

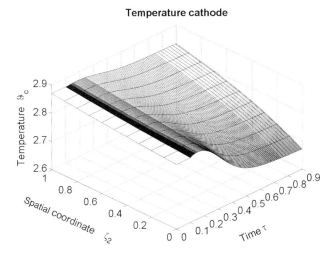

**Temperature cathode**

**Fig. 6.15** Short time behaviour of the cathodic gas temperature after a stepwise load change at $\tau = 0.1$. The states along a middle cathode channel are displayed.

So far, only the short time behaviour of the various states has been discussed. Concentrations and temperatures in the gas phases change within a time scale of a few seconds after which they reach a quasi steady state. In fact, there are processes in MCFC that are much slower than that. The temperature of the solid cell parts usually takes 1–2 h to come to a new steady state. Because many processes such as the electrochemical reaction rates and the heat exchange between gas phases and solid phase depend on the solid temperature, almost all other states in the gas phases continue to change until the solid temperature reaches steady state. Thus the dynamics of this state dominate the long time dynamics of the cell after a load change. Figure 6.16 shows the dynamic solid temperature along the right-hand side anode channel (running from $\zeta_1 = 0, \ldots, 1$ at $\zeta_2 = 0$). The temperature changes smoothly to moderately lower temperatures. This can be attributed to the lower cathode inlet temperature, which influences the solid temperature here. Figure 6.17 shows the temperature development on the opposite side of the fuel cell (left-hand side anode channel, at $\zeta_2 = 1$). Here, the minimum temperature at the anode outlet is slightly decreased, but the maximum temperature near the anode inlet is increased significantly. Not only is the hot spot increasing in temperature level, but it is also shifting from the corner of the cell ($\zeta_1 = 0$; $\zeta_2 = 1$) in the direction of the anode gas flow. This could be due to an increased cooling effect in the corner region due to higher reforming rates there. The high temperature difference in this simulation is not acceptable for the operation of a real cell system because it leads to mechanical stress and uneven catalyst degradation between hot and cold regions.

The different time constants of the MCFC can be demonstrated by plotting the cell voltage versus a logarithmic time axis. The cell voltage is influenced by many

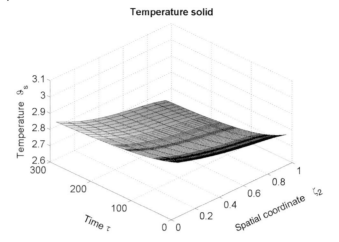

**Fig. 6.16** Long time behaviour of the solid temperature after a stepwise load change. The states along the right-hand-side anode channel are displayed.

states such as the gas concentrations and the solid temperature, so all the processes observed in the previous figures are observable here. Figure 6.18 shows that the cell voltage changes from its initial value to its new steady state in three steps. It only takes a few microseconds for the electrochemical double layers to change their charge density and thereby their potential differences. This is the virtually stepwise response on the load change observed in the current density dis-

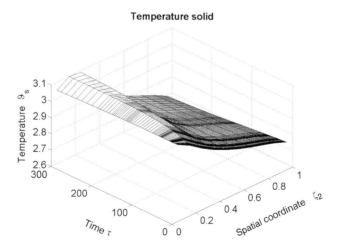

**Fig. 6.17** Long time behaviour of the solid temperature after a stepwise load change. The states along the left-hand-side anode channel are displayed.

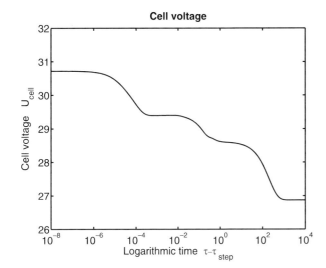

**Fig. 6.18** Dynamic behaviour of the cell voltage after a stepwise load change over a logarithmic time scale. Three dominant time constants become visible.

tribution. The capacities of the double layers, $c_a$, $c_e$ and $c_c$, are the time constants for this process. In a second step, the cell voltage changes in the next few seconds according to the gas concentrations and gas temperatures. The dimensionless volumes $V_a$, $V_m$ and $V_c$ are the responsible time constants. In the long time behaviour, the cell voltage changes together with the solid temperature. The large solid heat capacity, $c_{p,s}$, is the dominant time constant here. Regarding system control, it is practically impossible to control the fast processes. While it is difficult to control the processes with medium time constants, they have to be taken into account, because some of them produce inverse behaviour. For system control, the solid temperature is the most important state and it is governed by a large time constant at which the system can be controlled in MCFC.

## 6.4
## Summary

In this chapter, exemplary steady state and transient simulation results of the single MCFC reference model have been shown. The results indicate the location of hot temperature zones and the current density distribution within the cell area. Furthermore, they are fully interpretable and thus allow an improved understanding of the system behaviour. Due to the precedent parameter identification (see Chapter 5), it can also be applied to predict system states under varying operating conditions.

## Bibliography

**1** Heidebrecht, P., *Modelling, Analysis and Optimisation of a Molten Carbonate Fuel Cell with Direct Internal Reforming (DIR-MCFC)*, VDI Fortschritt-Berichte, Reihe 3, Nr. 826, VDI-Verlag, Düsseldorf, 2005.

**2** Heidebrecht, P., Sundmacher, K., Dynamic model of a cross-flow molten carbonate fuel cell with direct internal reforming (DIR-MCFC), *Journal of the Electrochemical Society* 152(1), 2005, A2217–A2228.

**3** Tränkle, F., Zeitz, M., Ginkel, M., Gilles, E.D., PROMOT: a modeling tool for chemical processes, *Mathematical and Computer Modelling of Dynamical Systems* 6(3), 2000, 283–307.

**4** Köhler, R., Mohl, K., Schramm, H., Zeitz, M., Kienle, A., Mangold, M., Stein, E., Gilles, E.D., in Vande Wouver, A., Saucez, P. and Schiesser, W. (Eds.), *Adaptive Method of Lines*, Chapman & Hall/CRC Press, Boca Raton, FL, 2001.

# 7
# Hot Spot Formation and Steady State Multiplicities

*Michael Krasnyk, Michael Mangold, Achim Kienle, and Kai Sundmacher*

## 7.1
## Introduction

Dynamic instabilities in chemical and electrochemical systems have been investigated over the last decades [4, 18]. It has been shown that in electrical and electrochemical systems multiplicities and oscillations can be caused by a negative differential resistance, i.e. by an electrical resistance decreasing with increasing current [7, 9, 14]. This property is found for quite different physical systems like gas discharge systems [19] or semiconductor devices [5, 17], and for various electrochemical reactions like the electro-oxidation of CO [6] or $H_2$ [21]. However, very little work has been done on nonlinear effects in fuel cells. Ertl and co-workers studied oscillations in the electrochemical methanol oxidation and pointed out a possible relevance for direct methanol fuel cells [10]. Datta and co-workers have investigated instabilities in proton exchange membrane fuel cells (PEMFCs) caused by the presence of CO in theoretical and experimental studies [22, 23]. Recently, Benziger and co-workers showed that in autohumidified PEMFCs bistabilities can occur due to an autocatalytic effect: Water formed in the electrochemical reaction increases the conductivity of the polymer membrane, hence accelerates the reaction, and rises the water concentration further, leading to a so-called 'wet spot' [1, 2, 15]. To our knowledge, there are hardly any publications available on instabilities and multiplicities in high temperature fuel cells.

The purpose of this chapter is to show that in high temperature fuel cells thermokinetic instabilities may result from the temperature-dependent electrical conductivity of the electrolyte. The effect can occur in solid oxide fuel cells (SOFCs) as well as in MCFCs. It is illustrated in Fig. 7.1. The charge transport in the electrolyte of a high temperature fuel cell is mainly accomplished by migration of ions. Therefore, the electrical conductivity of the electrolyte increases with increasing temperature (e.g. [16]). This property may be a potential source of instability: A local temperature rise reduces the resistance of the electrolyte and hence increases the local current density. As the current density is directly coupled to

*Molten Carbonate Fuel Cells.* Edited by Kai Sundmacher, Achim Kienle, Hans Josef Pesch, Joachim F. Berndt, and Gerhard Huppmann
Copyright © 2007 WILEY-VCH Verlag GmbH & Co. KGaA, Weinheim
ISBN: 978-3-527-31474-4

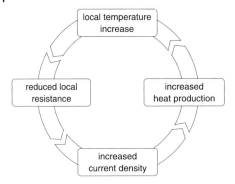

**Fig. 7.1** Temperature-dependent electrical conductivity causes thermal instabilities.

the reaction rate, the reaction rate and hence the reactive heat production increase locally. This causes a further temperature rise, i.e. the temperature disturbance is amplified. It will be shown later that this mechanism may narrow down the part of the electrolyte that actually transports charge. Channels of high current density may form in the electrolyte, and hot spots will result.

In this contribution, the possibility of hot spot formation is pointed out for various high temperature fuel cell models. In a first step, two model variants of a SOFC are investigated, a lumped model and a spatially distributed model. The intention is to show that hot spots can form under realistic operation conditions for reasonable model parameter values. A SOFC is chosen here instead of a MCFC, because it is much easier to find experimentally validated SOFC models in open literature with reliable data for kinetic parameters and for electrical conductivities (temperature-dependent conductivities for the electrolyte in the HotModule were not available). In the second part, a more general dimensionless high temperature fuel cell model is considered, which takes spatial dependences in one space coordinate into account. The results of this model will show that instabilities and hot spots may occur for a large range of kinetic and conductivity parameters.

When analysing the nonlinear behaviour of a fuel cell, one is always confronted with the problem that a stand-alone fuel cell is not of much technical use but should be studied in connection with some external electrical circuit. On the other hand, it is the objective of this paper to identify nonlinear effects actually caused by the phenomena inside the cell and not by the properties of an external device. To overcome this difficulty, this work considers three idealising modes of operation: (1) the case of perfect galvanostatic operation of the fuel cell, (2) the case of perfect potentiostatic operation, and (3) the case of a constant external ohmic resistance (see Fig. 7.2). The first two cases are taken from the way how fuel cells are characterised in a laboratory. In laboratory experiments, the fuel cell is often connected to a measuring instrument that either provides a constant current (galvanostat) or a constant voltage (potentiostat). This is achieved by control mechanisms inside the measuring instrument, which usually work so well

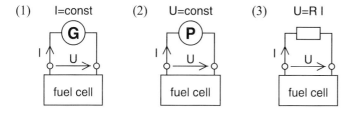

**Fig. 7.2** Operation modes of a fuel cell considered in this work:
(1) connection to a perfect galvanostat; (2) connection to perfect
potentiostat; (3) connection to an ohmic resistance.

that the cell current may be considered as an independent parameter for galvano-static operation and that the cell voltage may be considered as an independent parameter for potentiostatic operation. In the case of galvanostatic operation, the fuel cell responds to the fixed cell current with a certain cell voltage. In the case of potentiostatic operation, it responds to the fixed cell voltage with a certain cell current. The third case is a simple example of the use of the cell as an electrical power source. Here, the resistance of the external load is a free control parameter, while cell current and cell voltage are coupled by Ohm's law.

## 7.2
## Models for Nonlinear Analysis

### 7.2.1
### Spatially Distributed Model

Two SOFC models of different detailedness are used in the following. The first one is spatially one-dimensional and describes a counter-current cell as shown in Fig. 7.3. The complete list of model equations and parameter values can be found in the appendix. The main model assumptions are as follows:

The gases in the gas channels on anode and on cathode side behave ideal. A plug flow through the gas channels is considered. A pressure drop along the gas channels as well as pressure changes in time are neglected.

The mass transport through the electrodes to the reactive layer is limited by a mass transfer resistance. This is included in the model by a linear driving force approach.

The material transported through the porous electrodes to the reactive catalyst layer is consumed immediately in the electrochemical reactions, i.e. the electrodes' storage capacity for mass is negligible, because the electrodes are very thin.

The electrochemical reactions

$$H_2 + O^{2-} \rightarrow H_2O + 2e^-$$ (7.1)

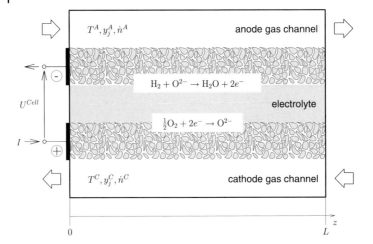

$T^A, y_j^A, \dot{n}^A$   anode gas channel

$H_2 + O^{2-} \rightarrow H_2O + 2e^-$

electrolyte

$\frac{1}{2}O_2 + 2e^- \rightarrow O^{2-}$

$U^{Cell}$

$I$

$T^C, y_j^C, \dot{n}^C$   cathode gas channel

0   $L$   $z$

**Fig. 7.3** Scheme of the counter-current FC considered in the spatially distributed model.

and

$$\frac{1}{2}O_2 + 2e^- \rightarrow O^{2-} \tag{7.2}$$

are assumed to take place on anode side and on cathode side, respectively. The anodic and cathodic reaction kinetics are of Butler–Volmer type and are taken from [3].

Transport of electrical charge is considered in the electrodes in the direction of the $z$-coordinate, and in the electrolyte in the direction perpendicular to the $z$-coordinate. The capacities of the double layers are neglected, as they are usually very small and as the interest of this paper focuses on the slower temperature dynamics of the cell.

Following [3], the conductivity of the electrodes is considered as invariant with temperature, whereas a temperature dependence of electrolyte's electrical conductivity is taken into account. The temperature dependence is described by an Arrhenius correlation. It is depicted in Fig. 7.4. Energy transport between the gaseous and the solid phases occurs due to convective heat transfer proportional to a heat transfer coefficient $\alpha$, and due to the enthalpy transport coupled to the mass fluxes. It is assumed that fluxes from the gas to the solid phase have the temperature of the gas and vice versa. On anode side, the transport of water from the anode to the anode gas channel has to be taken into account by a term in the temperature equation of the anode gas channel. Because there is no transport of material from the cathode to the cathode gas channel, a corresponding term is missing in the equation for the cathode gas temperature.

A pseudo-homogeneous energy balance is used for the anode, the cathode, and the electrolyte, assuming a high heat transfer and hence a vanishing temperature

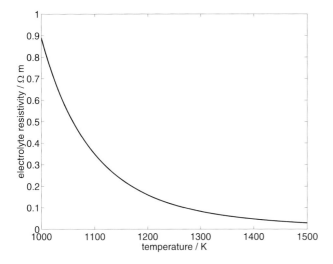

**Fig. 7.4** Temperature dependence of the electrolyte resistivity.

difference between the three solid phases. The resulting temperature equation for the pseudo-homogeneous solid phase contains source terms due to the heat of reaction, the conversion of internal energy into electrical energy, the heat generation due to ohmic losses in the electrodes, the heat transport by conduction in the solid phase, and the heat exchange between the gas channels and the solid parts of the cell.

The model is completed by an additional equation describing the external electrical circuit that depends on the considered case of operation.

### 7.2.2
### Lumped Model

The spatially distributed model is rather complex and requires a numerical analysis. Before carrying out such a numerical study, it seems worthwhile doing some qualitative considerations based on a simplified model. The simplified model results from the spatially distributed model by the following additional model assumptions: Spatial gradients are neglected, temperatures, concentrations, and potentials are constant across the whole cell area. The composition and temperature in the gas channels are considered as constant parameters, e.g. due to high flow rates in the channels. The mass transfer resistance between gas channels and solid as well as the electrical resistance of the metallic electrodes are neglected. The resulting model, which is listed in the appendix, consists of a temperature equation, of three implicit algebraic equations for the anodic and cathodic overpotentials and the total cell voltage, and of one additional equation describing the external electrical circuit.

**7.3**

**Analysis of the Lumped FC Model**

The purpose of this section is to determine conditions, where the simplified lumped model can possess multiple coexisting steady states. The easiest way to do this is by comparing the heat sinks and sources on the right-hand side of Eq. (7.40) for different cell temperatures – a method that has been in use for exothermic CSTRs for a long time [11, 20]. The steady state version of (7.40) can be written as

$$0 = Q_P(T^S) - Q_R(T^S) \tag{7.3}$$

$$Q_P(T^S) := \left( \frac{(-\Delta_R H)}{2F} - (\Phi^C - \Phi^A) \right.$$

$$\left. + \frac{c_{P,H_2}}{2F}(T^A - T^S) + \frac{c_{P,O_2}}{4F}(T^C - T^S) \right) \frac{I(T^S)}{LB} \tag{7.4}$$

$$Q_R(T^S) := 2\alpha T^S - \alpha(T^A + T^C) \tag{7.5}$$

The function $Q_R$ describes the removal of heat by exchange between the solid and the gas channels. It is linear in $T^S$ with a positive slope. The function $Q_P$ comprises all terms on the right-hand side of (7.40) that depend on $T^S$ in a nonlinear way and describes the heat production. The evaluation of $Q_P$ requires the solution of the implicit algebraic equations (7.37)–(7.39) in combination with one of Eqs. (7.34)–(7.36) for a given value of $T^S$. A steady state temperature is reached, if $Q_P$ equals $Q_R$. This condition can be evaluated graphically.

Figure 7.5 shows examples of this construction for different modes of operation of the cell. In case (1) of galvanostatic operation, the cell current is considered as a constant model parameter, and $Q_P$ is evaluated for varying values of $T^S$. It is found that only one intersection point is possible between the heat removal line $Q_R$, which increases with $T^S$, and the heat production curve $Q_P$, which decreases monotonically with $T^S$. Hence, the steady state is always unique for this model in the case of galvanostatic operation. The physical reason is that mainly the reaction kinetics contain the nonlinear terms of the model. Keeping the cell current constant means to fix the reaction rates of the cell and to suppress most of its nonlinear behaviour. However, it will be shown later that this conclusion holds for the lumped model only. In the spatially distributed model multiplicities can

**Fig. 7.5** Qualitative analysis of steady state multiplicities in the lumped FC model; (1) galvanostatic operation; (2) potentiostatic operation; (3) operation under a constant ohmic load; dotted line in (2) = $Q_P$ assuming a temperature independent electrolyte conductivity; white circles indicate intersections between heat removal line $Q_R$ and heat production curve $Q_P$, which define the steady state temperatures.

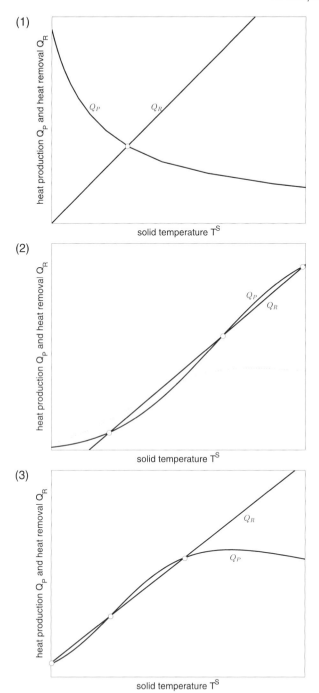

(1)

heat production $Q_P$ and heat removal $Q_R$

$Q_P$     $Q_R$

solid temperature $T^S$

(2)

heat production $Q_P$ and heat removal $Q_R$

$Q_P$   $Q_R$

solid temperature $T^S$

(3)

heat production $Q_P$ and heat removal $Q_R$

$Q_R$

$Q_P$

solid temperature $T^S$

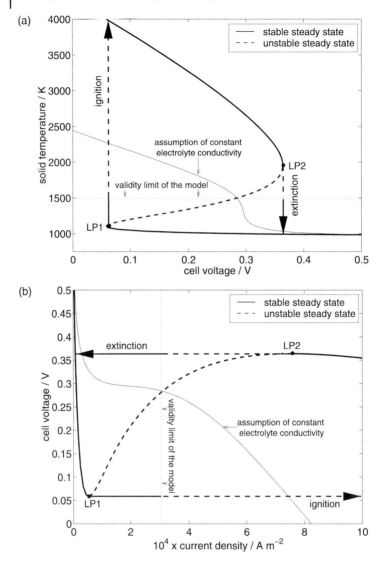

**Fig. 7.6** Bifurcation analysis of the lumped FC model for case (2) of potentiostatic operation; cell voltage $U^{Cell}$ is used as bifurcation parameter; $T^A = T^C = 980$ K; (a) solid temperature vs cell voltage; (b) cell voltage vs average current density $I/L/B$; thin lines in (a) and (b) show the system behaviour, if the electrolyte conductivity is assumed to be temperature independent.

also occur for galvanostatic operation, as this model possesses additional degrees of freedom.

In case (2) of potentiostatic operation, the cell voltage is kept constant, while the solid temperature $T^S$ is varied. The resulting heat production curve has the sigmoidal shape that is typical for exothermic reactors. The heat removal line can intersect this curve up to three times. Hence, up to three steady state solutions can coexist, of which the middle one is unstable, whereas the upper and the lower one are statically stable solutions. If the electrolyte's conductivity is assumed to be temperature independent (dotted line in the middle diagram of Fig. 7.5), then the heat production curve changes much less with temperature, and the range of model parameters where multiplicities can occur becomes much smaller. From this it can be concluded that the variance of the electrical conductivity with temperature plays a crucial role for the occurrence of multiple steady states in a FC.

The third case considered is the operation of the cell with a constant ohmic resistance (lower diagram of Fig. 7.5), where neither cell current nor cell voltage are independent parameters, but are coupled by Ohm's law. The qualitative behaviour in this operation mode is similar as under potentiostatic operation. Up to three steady states are found to coexist.

These qualitative results can be validated by a numerical bifurcation analysis of the lumped model (7.37)–(7.40) using some standard continuation package. Here, the continuation methods contained in the simulation tool DIVA [12] are applied. For potentiostatic operation, the cell voltage can be used as the independent bifurcation parameter. The model shows a hysteresis behaviour with a low-temperature branch of stable steady states and a coexisting stable high-temperature branch (Fig. 7.6). If the cell initially is operated at the low-temperature branch and the cell voltage is reduced below the value given by the limit point LP1, then the low temperature solution vanishes and the cell jumps to the high-temperature branch in a strong and sudden temperature rise. Obviously, the temperatures on the high-temperature solution branch are extremely high, above realistic operation temperatures of a FC and beyond the limit, where the model is still valid. However, the lower limit point LP1 lies in a region of reasonable operation conditions, i.e. the stability loss and the sudden temperature increase occur under conditions relevant for FC operation, even if the ignited steady state is never reached in reality. The characteristic cell voltage vs cell current in Fig. 7.6(b) has a rather unusual non-monotonous shape. This shape is caused by the temperature dependence of the electrolyte's electrical conductivity, as can be seen from a comparison to a model with constant conductivity. The thin lines in Fig. 7.6 show the results, if the electrolyte's conductivity is assumed to be constant at a value obtained from Eq. (7.30) for $T^S = 1500$ K. In this case, the cell voltage decreases monotonically with the cell current.

The operation of the cell under a ohmic load is studied by using the resistance of the external load as a bifurcation parameter. As can be seen in Fig. 7.7(a), a reduction of the load resistance below the value given by the lower limit point leads to sudden temperature increase and to the transient of the cell to a new steady

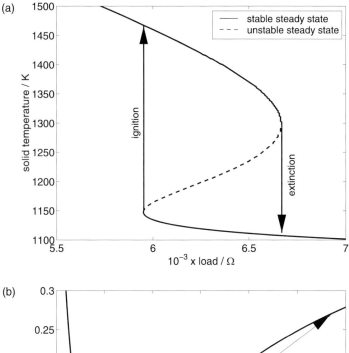

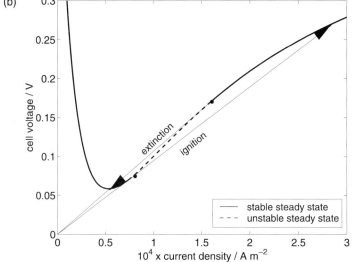

**Fig. 7.7** Bifurcation analysis of the lumped FC model for case (3) of operation under a constant ohmic load; the load resistance is used as bifurcation parameter; $T^A = T^C = 980$ K; (a) solid temperature vs cell voltage; (b) cell voltage vs average current density $I/L/B$.

state on the high temperature branch. A subsequent increase of the load resistance above the value given by the upper limit point results in a drop of the cell temperature to the low-temperature branch of steady state solutions. The corresponding voltage vs current characteristic in Fig. 7.7(b) has exactly the same shape as for the potentiostatic operation, Fig. 7.6(b). Coupling the cell voltage

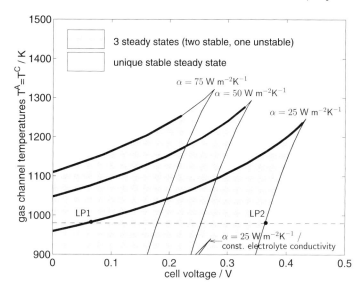

**Fig. 7.8** Multiplicity regions of the lumped FC model for the case of potentiostatic operation; bold lines indicate limit points with a solid temperature below 1500 K.

and cell current via an ohmic resistor only affects the stability of the solutions. In Fig. 7.8, additional model parameters and operation conditions are included in the analysis of multiple steady states. Potentiostatic operation is considered. The figure shows multiplicity regions in a parameter plane defined by the operation parameters cell voltage and gas temperature. Operation conditions, where multiple steady states exist, lie within the shaded areas in the figure. The boundaries of the shaded areas are formed by limit points. For example, the limit points LP1 and LP2 from Fig. 7.6 are found in Fig. 7.8 as intersection points of the dashed line $T^A = T^C = 980$ K and of the boundary of the multiplicity region for $\alpha = 25$ W m$^{-2}$ K$^{-1}$. If the gas temperatures are increased, then the limit points and the region of steady states move towards higher values of the cell voltage. At the same time, the distance between the limit points decreases, i.e. the range of values of the cell voltage, where multiplicities exist, becomes smaller for higher gas temperatures. In Fig. 7.8, the two limit points coincide for $\alpha = 25$ W m$^{-2}$ K$^{-1}$ at gas temperatures of about 1250 K and a cell voltage of about 0.4 V. For gas temperatures above the value of 1250 K, the multiplicities vanish completely. Figure 7.8 also shows that the heat transfer coefficient between gas and solid phases $\alpha$ only affects the quantitative behaviour of the model, but hardly the qualitative results. Increasing the heat transfer coefficient only shifts the multiplicity region towards smaller values of the cell voltage and towards higher values of the gas temperatures, but cannot destroy it. However, the result changes strongly, if a temperature independent conductivity of the electrolyte is assumed. In this case, the multiplicity region shrinks to a tiny area in Fig. 7.8, indicating again that the occurrence of multiple steady states is mainly caused by the varying electrolyte conductivity.

**7.4**
**Analysis of the Spatially Distributed FC Model**

The investigations of the previous section are now extended to the spatially distributed model variant of the FC. Additional phenomena taken into account in this more detailed model are the varying composition and temperature of the anode and cathode gases along the flow channels, the mass transfer resistance between gas and solid phases, and the spatial dependence of the electrical potentials and the solid temperature. In order to get a spatially discretised system with differential index one, the time derivative of the total concentration in (7.15) is eliminated by using the thermal equation of state (7.22) and the temperature equations (7.17) and (7.19).

Figure 7.9 shows multiplicity regions of the distributed model under potentiostatic operation and can be seen in connection with results for the lumped model in Fig. 7.8. Because the temperature of the gas channels are no more model parameters in the detailed model, the inlet temperatures of the gases are treated as the second bifurcation parameter instead. The resulting multiplicity region is similar to the multiplicity region of the lumped model. However, the additional degrees of freedom of the spatially distributed model are reflected by the coexistence of up to five steady states, whereas in the lumped model only three coexisting steady states are found. Figure 7.10 gives a closer look on the multiplicity region. It was obtained by varying the cell voltage under potentiostatic operation for constant inlet gas temperatures of 1000 K, i.e. by moving along the lower border of Fig. 7.9.

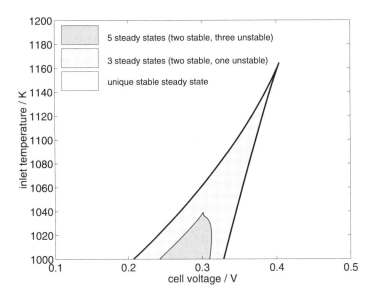

**Fig. 7.9** Multiplicity regions of the spatially distributed FC model for case (2) of potentiostatic operation.

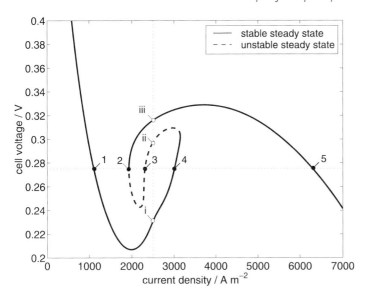

**Fig. 7.10** Cell voltage vs average current density $I/L/B$ of the spatially distributed FC model ($T_{in}^A = T_{in}^C = 1000$ K); 1–5 are coexisting steady states under potentiostatic operation; i–iii are coexisting steady states under galvanostatic operation; stability information in the figure refers to galvanostatic operation.

One can see that for a given cell voltage of, e.g. 0.28 V the cell can be in five different steady states 1–5 with different cell temperatures and different cell currents. Further, the figure shows that now multiple steady states can also coexist for a given cell current, e.g. states i,ii, and iii for a average current density of 2500 A m$^{-2}$, whereas the steady states of the lumped model under galvanostatic operation are always unique. The reason for this difference is that galvanostatic operation of the distributed model only fixes the overall cell current, but allows differences in the spatial distribution of the current density that are not taken into account in the lumped model. The physical mechanism behind the galvanostatic multiplicities is the interaction between the heat production by the electrochemical reaction and the temperature dependence of the electrolyte's electrical conductivity. This interaction may narrow down the charge transport through the electrolyte to a small portion of the electrode area, resulting in a channel of high current density and a hot spot. The coexisting spatial profiles of the temperature and the current density in Fig. 7.11 illustrate this effect. In the case of solution i, the current density along the space coordinate varies comparatively weakly, and the maximum temperature is rather low. In the case of solution iii, the electrochemical reaction and the production of heat and electrical current concentrates on the left boundary region of the cell, where a high temperature peak forms. The unstable solution ii lies in the middle between the two stable solutions. A transient from solution i to solution iii can be reached by a sufficiently

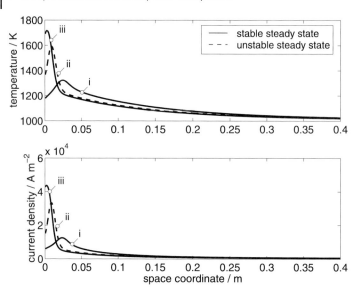

**Fig. 7.11** Coexisting spatial profiles of the temperature and the current density of the distributed model under case (1) of galvanostatic operation Im ($T_{in}^A = T_{in}^C = 1000$ K, $I/L/B = 2500$ A m$^{-2}$).

strong local disturbance of the cell temperature. A local temperature increase reduces the local resistivity of the electrolyte and hence increases the local current density and the local reaction rate. This leads to an increase of the local heat production, to a further temperature rise, and to a further reduction of the local resistivity. This destabilising effect is counteracted by the heat transport in axial direction due to heat conduction. A new equilibrium, solution iii, is reached when the temperature gradients are sufficiently steep that enough heat can be removed from the hot zone by conduction.

The coexistence of the spatially more homogeneous solution of type i and the hot spot solution of type iii is limited to a certain range of average cell current densities as follows from Fig. 7.10. Under the conditions shown in Fig. 7.10 the type i solution is unique for cell current densities below 2000 A m$^{-2}$, whereas for current densities above 3500 A m$^{-2}$ solely the hot spot solution of type iii exists.

## 7.5
### Analysis of a More General High Temperature Fuel Cell Model

In the following, the results of the previous sections are extended to a more general fuel cell model that can describe a FC or MCFC, depending on the values chosen for the kinetic parameters. The intention is to identify a range of parameter values, for which multiplicities and hot spots may occur. As the main interest is in the influence of the electrolyte's conductivity, the model is simplified further compared to the SOFC model of this chapter, or compared to the models of the HotModule presented in Chapter 3. For simplicity, it is assumed that the composition of the gases in the anode and cathode gas channels does not change. This assumption is realistic for high flow rates of the gases and low fuel utilisation. As a direct consequence of this assumption, the reaction rates are taken to be of zeroth order in the reactants, i.e. concentration polarisation is neglected. The resulting model equations read in dimensionless form

$$\frac{\partial \Theta}{\partial \tau} = \frac{\partial^2 \Theta}{\partial \eta^2} + (B - \phi^{tot})i' - Bi_1\Theta \tag{7.6}$$

$$\frac{\partial \Theta}{\partial \eta}\bigg|_{0,\tau} = Bi_2\Theta(0, \tau) \tag{7.7}$$

$$\frac{\partial \Theta}{\partial \eta}\bigg|_{1,\tau} = -Bi_2\Theta(1, \tau) \tag{7.8}$$

$$i' = \psi^A \exp\left(\gamma^A \frac{\Theta}{1+\Theta}\right)\left\{\exp\left(-(1-\beta^A)\gamma^{eq}\frac{\phi^A}{1+\Theta}\right)\right.$$
$$\left. - K_{eq}^A \exp\left(\beta^A\gamma^{eq}\frac{\phi^A}{1+\Theta}\right)\right\} \tag{7.9}$$

$$i' = \psi^C \exp\left(\gamma^C \frac{\Theta}{1+\Theta}\right)\left\{\exp\left(-(1-\beta^C)\gamma^{eq}\frac{\phi^C}{1+\Theta}\right)\right.$$
$$\left. - K_{eq}^C \exp\left(\beta^C\gamma^{eq}\frac{\phi^C}{1+\Theta}\right)\right\} \tag{7.10}$$

$$i' = \psi^E \exp\left(\gamma^E \frac{\Theta}{1+\Theta}\right)(\phi^A + \phi^C - \phi^{tot}) \tag{7.11}$$

$$I = \int_0^1 i' \, d\eta \tag{7.12}$$

For details of the derivation of this model, see [13]. Equation (7.6) has the structure of a reaction diffusion equation with a nonlinear source that is given implicitly by Eqs. (7.9)–(7.12). The space integral in the charge balance (7.12) complicates the analysis a bit. The determination of steady state solutions is rather simple for the potentiostatic operation mode, because in this case (7.12) becomes an explicit equation for the total cell current $I$. Then the steady state behaviour can be investigated semi-analytically by constructing phase portraits, as was shown in [13]. Under galvanostatic operation, $I$ is a constant parameter, and (7.12) can be seen as an implicit equation for $\phi^{tot}$. The analysis of this case requires numerical analysis. A numerical bifurcation analysis of the model (7.6)–(7.12) was done using a recently developed software tool [8] for the continuation of singularities. The diagrams in Fig. 7.12 show the result of a bifurcation analysis, when galvanostatic operation is assumed and $I$ is used as the bifurcation parameter. The central diagram contains the occurring co-dimension 1 singularities

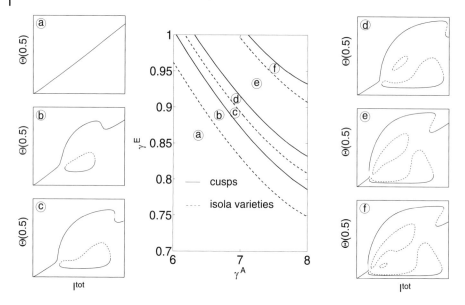

**Fig. 7.12** Bifurcation analysis of the model (7.6)–(7.12) under galvanostatic operation; center diagram: isola and hysteresis varieties in the $\gamma^A$–$\gamma^E$ parameter plane; surrounding diagrams (a)–(f): typical bifurcation diagrams for the six regions of the centre diagram.

in a parameter plane spanned by the activation energy of the anodic reaction $\gamma^A$ and the activation energy of the electrolyte's conductivity $\gamma^E$. The singularities separate the parameter plane into regions with qualitatively different behaviour. Typical one-parameter continuation plots for each of the region are depicted in the surrounding diagrams. In region (a), only one stable steady state exists. In region (b), there is an isolated solution branch in addition to the stable solution branch that also exists in region (a). Consequently, up to three coexisting steady state solutions can be found in region (b). In region (c), the formerly completely stable solution branch is folded to a hysteresis for certain values of $I$. In region (d) to (f), further isolas and hystereses appear, until in region (f) up to seven steady states coexist. Figure 7.13 gives the corresponding temperature profiles. Although the temperature maxima of different solutions vary strongly, the total cell voltages of all coexisting solutions are found to be very close to each other. This means that at a given cell current the same electrical power may be achieved at quite different temperature levels, which has consequences for the operation of a high temperature fuel cell. The temperature maxima of the stable solutions (1) and (7) in Fig. 7.13 are very high and inconvenient for the cell operation. The simplest way to avoid them is to keep the cell current below the threshold given by the utter left hysteresis in Fig. 7.12(f). However, this measure also cuts down the power density of the cell. It seems much more attractive to make use of the unstable solutions (4) and (5) in Fig. 7.13. This would require a simple feedback

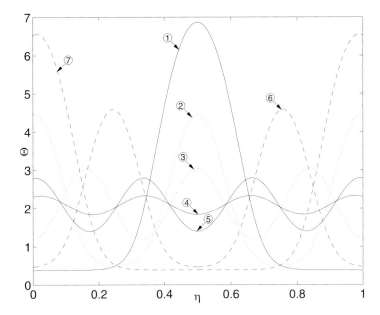

**Fig. 7.13** Coexisting spatial profiles of the temperature under the conditions of Fig. 7.12(f).

control for stabilization, but the additional effort is rewarded by a rather even temperature profile and a high power density.

## 7.6
## Conclusions

Temperature control is a crucial point for the operation of high temperature fuel cells. The prediction and avoidance of local over temperatures requires a thorough understanding of the various processes of heat production and heat transport interacting in a fuel cell. From the analysis of classical chemical fixed bed reactors, a number of mechanisms are known to be responsible for the formation of hot spots, e.g. the decrease of the feed temperature or the interaction between exothermic and endothermic reactions. These mechanisms are also relevant for high temperature fuel cells. However, this work indicates that in high temperature fuel cells an additional effect has to be taken into account that is not present in conventional fixed bed reactors. This effect is the temperature dependence of the electrical conductivity of the electrolyte. The existence of multiplicities depends mainly on the qualitative property of a negative differential resistance of the system and less on the quantitative type of the temperature dependence.

This study considers three modes of operation of the fuel cell: the connection to a perfect galvanostat, to a perfect potentiostat, and to an ohmic resistance. If spatial gradients in the fuel cell are neglected, multiple steady states exist only in the last two cases. However, if a spatially distributed model is used, then coexisting steady states are also found for a fixed cell current. The reason for this behaviour is that a fixed cell current only determines the total heat generated by the chemical reaction, but allows for degrees of freedom with respect to the spatial distribution of the heat generation.

## 7.7
## Appendix: Model Equations for Nonlinear Analysis

### 7.7.1
### Equations of the Spatially Distributed Model

- Anode and cathode gas channels:
  - Component material balances:

$$\frac{\partial}{\partial t}(y_j^{A/C} c_t^{A/C}) = \mp \frac{1}{H^{A/C} B} \frac{\partial \dot{n}_j^{A/C}}{\partial z} + \frac{v_j^{A/C}}{H^{A/C}} \frac{i}{2F} \qquad (7.13)$$

$$y_j^A(0, t) = y_{j,\text{in}}^A(t), \quad y_j^C(L, t) = y_{j,\text{in}}^C(t) \qquad (7.14)$$

  ($j = H_2, H_2O$ on anode side, and $j = N_2, O_2$ on cathode side)
  - Total material balance:

$$\frac{\partial c_t^{A/C}}{\partial t} = \mp \frac{1}{H^{A/C} B} \frac{\partial \dot{n}^{A/C}}{\partial z} + \frac{1}{H^{A/C}} \sum_j v_j^{A/C} \frac{i}{2F} \qquad (7.15)$$

$$\dot{n}^A(0, t) = \dot{n}_{\text{in}}^A(t), \quad \dot{n}^C(L, t) = \dot{n}_{\text{in}}^C(t) \qquad (7.16)$$

  - Temperature equation on anode side:

$$c_t^A c_P^A \frac{\partial T^A}{\partial t} = -\frac{\dot{n}^A}{H^A B} c_P^A \frac{\partial T^A}{\partial z} + \frac{i}{2F} \frac{c_{P,H_2O}}{H^A}(T^S - T^A)$$

$$+ \frac{\alpha}{H^A}(T^S - T^A) \qquad (7.17)$$

$$T^A(0, t) = T_{\text{in}}^A(t) \qquad (7.18)$$

  Temperature equation on cathode side:

$$c_t^C c_P^C \frac{\partial T^C}{\partial t} = \frac{\dot{n}^C}{H^C B} c_P^C \frac{\partial T^C}{\partial z} + \frac{\alpha}{H^C}(T^S - T^C) \qquad (7.19)$$

$$T^C(L, t) = T_{\text{in}}^C(t) \qquad (7.20)$$

Specific molar heat capacity:

$$c_P^{A/C} = \sum_j y_j^{A/C} c_{P,j}$$ (7.21)

– Thermal equation of state:

$$c_t^{A/C} = \frac{p}{\mathbb{R} T^{A/C}}$$ (7.22)

• Solid phase:
  – Component material balances:

$$0 = D_{\text{eff}}^{A/C} c_t \frac{y_j^{A/C} - y_j^S}{d^{A/C}} + v_j^{A/C} \frac{i}{2F}$$ (7.23)

  – Anodic reaction kinetics:

$$i = \gamma^A y_{H_2}^S y_{H_2O}^S \exp\left(-\frac{E^A}{\mathbb{R} T^S}\right)$$
$$\times \left\{ \exp\left(\theta_a^A \frac{F}{\mathbb{R} T^S} \eta^A\right) - \exp\left(-\theta_c^A \frac{F}{\mathbb{R} T^S} \eta^A\right) \right\}$$ (7.24)

  – Cathodic reaction kinetics:

$$i = \gamma^C y_{O_2}^{S\,0.25} \exp\left(-\frac{E^C}{\mathbb{R} T^S}\right)$$
$$\times \left\{ \exp\left(\theta_a^C \frac{F}{\mathbb{R} T^S} \eta^C\right) - \exp\left(-\theta_c^C \frac{F}{\mathbb{R} T^S} \eta^C\right) \right\}$$ (7.25)

  – Charge balances in the electrodes:

$$\frac{\partial}{\partial z}\left(\frac{d^{A/C}}{\rho^{A/C}} \frac{\partial \Phi^{A/C}}{\partial z}\right) = \pm i$$ (7.26)

$$\Phi^A(0,t) = 0, \quad \frac{B d^C}{\rho^C} \left.\frac{\partial \Phi^C}{\partial z}\right|_{0,t} = I,$$
$$\left.\frac{\partial \Phi^A}{\partial z}\right|_{L,t} = \left.\frac{\partial \Phi^C}{\partial z}\right|_{L,t} = 0$$ (7.27)

  – Voltage drop in the electrolyte:

$$U^{\text{Cell}} = \Phi^C(0,t) - \Phi^A(0,t)$$
$$= U^0(T^S) - \eta^A - \eta^C - \rho^E(T^S) d^E i$$ (7.28)

- Open circuit voltage:

$$U^0(T^S) = -\frac{1}{2F}\left(\Delta_R G - \Delta_R S(T^S - T_{\text{ref}})\right.$$

$$\left. + \mathbb{R}T^S \ln \frac{y_{H_2O}^A}{y_{H_2}^A y_{O_2}^{C\ 0.5}}\right) \tag{7.29}$$

- Electrical conductivity of the electrolyte:

$$\rho^E = \frac{1}{\beta_1}\exp\left(\frac{\beta_2}{T^S}\right) \tag{7.30}$$

- Temperature equation for the solid phase:

$$(d^A + d^E + d^C)(\rho c_P)^S\frac{\partial T^S}{\partial t} = \left(\frac{(-\Delta_R H)}{2F} - (\Phi^C - \Phi^A)\right)i$$

$$+ \frac{d^A}{\rho^A}\left(\frac{\partial \Phi^A}{\partial z}\right)^2 + \frac{d^C}{\rho^C}\left(\frac{\partial \Phi^C}{\partial z}\right)^2 + (d^A + d^E + d^C)\lambda\frac{\partial^2 T^S}{\partial z^2}$$

$$+ \left(\alpha + \frac{i}{2F}c_{P,\,H_2}\right)(T^A - T^S)$$

$$+ \left(\alpha + \frac{i}{4F}c_{P,\,O_2}\right)(T^C - T^S) \tag{7.31}$$

$$-\lambda\frac{\partial T^S}{\partial z}\bigg|_{0,\,t} = \alpha(T_{\text{amb}} - T^S(0, t)) \tag{7.32}$$

$$\lambda\frac{\partial T^S}{\partial z}\bigg|_{L,\,t} = \alpha(T_{\text{amb}} - T^S(L, t)) \tag{7.33}$$

- Depending on the mode of operation, one of the following three equations is used for describing the external electrical circuit:
  - Case (1): galvanostatic operation

$$I = constant \tag{7.34}$$

  - Case (2): potentiostatic operation

$$U^{\text{Cell}} = constant \tag{7.35}$$

  - Case (3): operation with an external ohmic load

$$U^{\text{Cell}} = RI \tag{7.36}$$

7.7.2
**Equations of the Lumped Model**

- Anodic reaction kinetics:

$$\frac{I}{LB} = \gamma^A y_{H_2}^S y_{H_2O}^S \exp\left(-\frac{E^A}{\mathbb{R}T^S}\right)$$

$$\times \left\{ \exp\left(\theta_a^A \frac{F}{\mathbb{R}T^S} \eta^A\right) - \exp\left(-\theta_c^A \frac{F}{\mathbb{R}T^S} \eta^A\right) \right\} \tag{7.37}$$

- Cathodic reaction kinetics:

$$\frac{I}{LB} = \gamma^C y_{O_2}^{S~0.25} \exp\left(-\frac{E^C}{\mathbb{R}T^S}\right)$$

$$\times \left\{ \exp\left(\theta_a^C \frac{F}{\mathbb{R}T^S} \eta^C\right) - \exp\left(-\theta_c^C \frac{F}{\mathbb{R}T^S} \eta^C\right) \right\} \tag{7.38}$$

- Voltage drop in the electrolyte:

$$U^{Cell} = \Phi^C - \Phi^A$$

$$= U^0(T^S) - \eta^A - \eta^C - \rho^E(T^S)d^E \frac{I}{LB} \tag{7.39}$$

- Temperature equation for the solid phase:

$$(d^A + d^E + d^C)(\rho c_P)^S \frac{dT^S}{dt} = \left(\frac{(-\Delta_R H)}{2F} - (\Phi^C - \Phi^A)\right)\frac{I}{LB}$$

$$+ \left(\alpha + \frac{c_{P,H_2}}{2F} \frac{I}{LB}\right)(T^A - T^S)$$

$$+ \left(\alpha + \frac{c_{P,O_2}}{4F} \frac{I}{LB}\right)(T^C - T^S) \tag{7.40}$$

- External load described by Eqs. (7.34), (7.35), or (7.36).

### 7.3.3
**Model Parameters**

| | |
|---|---|
| $B = 0.04$ m[a] | $\Delta_R S = -57$ J mol$^{-1}$ K$^{-1}$ |
| $c_{P,\,H_2} = 30$ J mol$^{-1}$ K$^{-1}$ | $T_{in}^{A/C} = 1000$ K |
| $c_{P,\,H_2O} = 44$ J mol$^{-1}$ K$^{-1}$ | $T_{ref} = 1300$ K |
| $c_{P,\,N_2} = 34$ J mol$^{-1}$ K$^{-1}$ | $y_{H_2,\,in}^A = 0.9$ |
| $c_{P,\,O_2} = 36$ J mol$^{-1}$ K$^{-1}$ | $y_{H_2O,\,in}^A = 0.1$ |
| $d^A = 10^{-4}$ m | $y_{N_2,\,in}^C = 0.8$ |
| $d^C = 0.5 \times 10^{-4}$ m[a] | $y_{O_2,\,in}^C = 0.2$ |
| $d^E = 1.8 \times 10^{-4}$ m[a] | $\alpha = 25$ Wm$^{-2}$ K$^{-1}$ |
| $D_{eff}^A = 3.5 \times 10^{-5}$ m$^2$ s$^{-1}$[a] | $\beta_1 = 3.34 \times 10^{-4}$ $\Omega^{-1}$ m$^{-1}$[a] |
| $D_{eff}^C = 7.3 \times 10^{-6}$ m$^2$ s$^{-1}$[a] | $\beta_2 = 1.03 \times 10^4$ K[a] |
| $E^A = 140{,}000$ J mol$^{-1}$[a] | $\gamma^A = 5.7 \times 10^7$ A m$^{-2}$[a] |
| $\Delta_R G = -175{,}933$ J mol$^{-1}$ | $\gamma^C = 7 \times 10^9$ A m$^{-2}$[a] |
| $E^C = 160{,}000$ J mol$^{-1}$[a] | $\theta_a^A = 2$[a] |
| $H^A = 10^{-3}$ m | $\theta_c^A = 1$[a] |
| $H^C = 10^{-3}$ m | $\theta_a^C = 1.4$[a] |
| $\Delta_R H = -241{,}830$ J mol$^{-1}$ | $\theta_c^C = 0.6$[a] |
| $L = 0.4$ m | $\lambda = 0.7$ W m$^{-1}$ K$^{-1}$ |
| $\dot{n}_{in}^A = 1.39 \times 10^{-3}$ mol s$^{-1}$[a] | $\rho^A = 1.5 \times 10^{-7}$ $\Omega$ m[a] |
| $\dot{n}_{in}^C = 3.8 \times 10^{-2}$ mol s$^{-1}$[a] | $\rho^C = 4.2 \times 10^{-8}$ $\Omega$ m[a] |
| $p = 100{,}000$ Pa | $(\rho c_P)^S = 10^6$ J m$^{-3}$ K$^{-1}$ |

[a] source: [3]

### Bibliography

**1** J.B. Benziger, E.J. Chia, E. Karnas, J. Moxley, C. Teuscher, and I.G. Kevrekidis. The stirred tank reactor polymer electrolyte membrane fuel cell. *AIChE Journal*, 50:1889–1900, 2004.

**2** E.J. Chia, J.B. Benziger, and I.G. Kevrekidis. Water balance and multiplicity in a polymer electrolyte membrane fuel cell. *AIChE Journal*, 50:2320–2324, 2004.

**3** P. Costamagna and K. Honegger. Modeling of solid oxide heat exchanger integrated stacks and simulation at high fuel utilization. *Journal of the Electrochemical Society*, 145:3995–4007, 1998.

**4** J.L. Hudson and T.T. Tsotsis. Electrochemical reaction dynamics: a review. *Chemical Engineering Science*, 49:1493–1572, 1994.

**5** G. Hüpper, E. Schöll, and L. Reggiani. Global bifurcation and hysteresis of self-generated oscillations in a microscopic model of nonlinear transport in p-Ge. *Solid-State Electronics*, 32:1787–1791, 1989.

**6** M.T.M. Koper, T.J. Schmidt, N.M. Marković, and P.N. Ross. Potential oscillations and S-shaped polarization curve in the continuous electro-oxidation of CO on platinum single-crystal electrodes. *Journal of Physical Chemistry B*, 105:8381–8386, 2001.

**7** M.T.M. Koper and J.H. Sluyters. Instabilities and oscillations in simple models of electrocatalytic surface reactions. *Journal of Electroanalytical Chemistry*, 371:149–159, 1994.

**8** M. Krasnyk, M. Ginkel, M. Mangold, and A. Kienle. Numerical analysis of higher-order singularities in complex chemical process models in ProMoT. In L. Puigjaner and A. Espuna, editors, *European Symposium on Computer Aided Process Engineering – 15*, pp. 223–228, 2005, Elsevier, Amsterdam.

9  K. Krischer, N. Mazouz, and G. Flätgen. Pattern formation in globally coupled electrochemical systems with an S-shaped current-potential curve. *Journal of Physical Chemistry B*, 104:7545–7553, 2000.

10  J. Lee, C. Eickes, M. Eiswirth, and G. Ertl. Electrochemical oscillations in the methanol oxidation on Pt. *Electrochimica Acta*, 47:2297–2301, 2002.

11  F.G. Liljenroth. Starting and stability phenomena of ammonia-oxidation and similar reactions. *Chemical and Metallurgical Engineering*, 19:287–293, 1918.

12  M. Mangold, A. Kienle, K.D. Mohl, and E.D. Gilles. Nonlinear computation in DIVA – methods and applications. *Chemical Engineering Science*, 55:441–454, 2000.

13  M. Mangold, M. Krasnyk, and K. Sundmacher. Nonlinear analysis of current instabilities in high temperature fuel cells. *Chemical Engineering Science*, 59:4869–4877, 2004.

14  N. Mazouz and K. Krischer. A theoretical study on Turing patterns in electrochemical systems. *Journal of Physical Chemistry B*, 104:6081–6090, 2000.

15  J.F. Moxley, S. Tulyani, and J.B. Benziger. Steady state multiplicity in the autohumidification polymer electrolyte membrane fuel cell. *Chemical Engineering Science*, 58:4705–4708, 2003.

16  B. Munder, Y. Ye, L. Rihko-Struckmann, and K. Sundmacher. Solid electrolyte membrane reactor for controlled partial oxidation of hydrocarbons: Model and experimental validation. *Catalysis Today*, 104:138–148, 2004.

17  F.-J. Niedernostheide, editor. *Nonlinear Dynamics and Pattern Formation in Semiconductors and Devices*. Springer, Berlin, 1995.

18  L.M. Pismen. Kinetic instabilities in man-made and natural reactors. *Chemical Engineering Science*, 35:1950–1978, 1980.

19  H.-G. Purwins, C. Radehaus, T. Dirksmeyer, R. Dohmen, R. Schmeling, and H. Willebrand. Pattern formation in gas discharge systems. *Physics Letters A*, 125:92–94, 1987.

20  C. van Heerden. The character of the stationary state of exothermic processes. *Chemical Engineering Science*, 7:133–145, 1958.

21  H. Varela and K. Krischer. Nonlinear phenomena during electrochemical oxidation of hydrogen on platinum electrodes. *Catalysis Today*, 70:411–425, 2001.

22  J. Zhang and R. Datta. Higher power output in a PEMFC operating under autonomous oscillatory conditions in the presence of CO. *Electrochemical and Solid State Letters*, 7:A37–A40, 2004.

23  J. Zhang, J.D. Fehribach, and R. Datta. Mechanistic and bifurcation analysis of anode potential oscillations in PEMFCs with CO in anode feed. *Journal of the Electrochemical Society*, 151:A689–A697, 2004.

# 8
# Conceptual Design and Reforming Concepts

*Peter Heidebrecht and Kai Sundmacher*

The reference model introduced in Chapter 3 is a spatially distributed transient models with a high level of detail. It simulates a MCFC single cell with good precision and is used for simulation, system analysis, development of control strategies and system optimisation. But it cannot give clear and evident insight into the basic interaction of the competing processes in this system. In the direct internal reforming concept in MCFC, the reforming process is intensively coupled to the electrochemical reactions at the anode. The substantial coupling allows to overcome thermodynamic equilibrium limitations of the reforming process, and the thermal exchange allows the endothermic reforming process to be heated by the heat producing oxidation. While these certainly are strong advantages, high temperature fuel cells offer other reforming concepts as well (see Fig. 8.1). To analyse the interplay of reforming and oxidation on a conceptual level, a new model is required which focuses on the steady state behaviour of only the anode. It should be

- physically meaningful in order to allow for an intuitive understanding of its results,
- flexible, so that it can easily be extended or modified for other purposes,

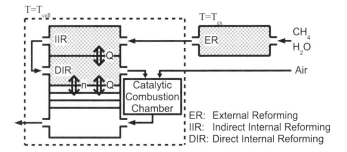

**Fig. 8.1** Reforming concepts for high temperature fuel cells.

*Molten Carbonate Fuel Cells.* Edited by Kai Sundmacher, Achim Kienle, Hans Josef Pesch, Joachim F. Berndt, and Gerhard Huppmann
Copyright © 2007 WILEY-VCH Verlag GmbH & Co. KGaA, Weinheim
ISBN: 978-3-527-31474-4

- fast to solve, as the task here is not a detailed and exact prediction of states, but should rather yield principal insight and show up tendencies,
- easily interpretable and illustrative,
- comparable to the reference model so that the results can be transferred between both models.

The steady state anode model fulfills these requirements. It is directly derived from the reference model, but it contains a much smaller number of state variables and considers only the most important dependences. In the following, the assumptions and the equations for the steady state anode model are given. After that, the conversion diagram is introduced, which is the illustration tool for this model. The various applications of this model are discussed afterwards. For derivations and detailed discussions we refer to other publications [1, 2].

## 8.1
## Steady State Anode Model

### 8.1.1
### General

The steady state anode model is derived under the following assumptions:
- The system is at steady state.
- A sufficiently high cathode gas recycle ratio is assumed, so that the composition and temperature of the cathode gas are approximately constant along the channel.
- As a consequence of the constant cathode gas temperature, the cell temperature is also constant along the spacial coordinate. Because this temperature can be adjusted by the amount of air input into the system, it is considered a parameter. Nevertheless, temperature effects on the reaction rates and the equilibrium constants are considered.
- The oxidation reaction of carbon monoxide (Eq. (3.4)) in MCFC is known to be of minor importance compared to the oxidation of hydrogen and is thus neglected.
- Due to the very fast water–gas shift reaction and because the equilibrium of this reaction at high temperatures is on the right-hand side of Eq. (3.2), the amount of carbon monoxide in the anode gas is negligible. Thus, both reforming reactions are condensed into one single reforming reaction:

$$CH_4 + 2H_2O \rightleftharpoons CO_2 + 4H_2 \qquad (8.1)$$

- Mass transport resistance in the electrode pores is neglected.

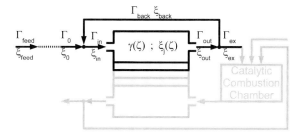

**Fig. 8.2** Reforming concepts for high temperature fuel cells.

- Instead of galvanostatic cell operation, potentiostatic operating mode is applied. Thus, the cell voltage is given as an operating parameter.
- The ion conduction resistance in the electrolyte perpendicular to the cell plain is negligible. Nevertheless, no ion conduction occurs along the cell plain. Thus the electrolyte has identical electric potentials at the anode and cathode electrode, but these are spacially distributed.
- The possibility of anode gas recycling shall be considered (see Fig. 8.2).

To reduce the number of state variables further, a physically motivated transformation is applied. The idea is that for a given composition of the feed gas ($CH_4$/ $H_2O$-mixture, characterised by the steam-to-carbon ratio) the composition of the gas at any point in the anode channel or in any of the reforming units is described by only two states: the extent of the reforming reaction, assigned $\xi_{ref}$, and the extent of the oxidation reaction, assigned $\xi_{ox}$. These variables are made dimensionless and normalised to unity, so they can be interpreted as follows:

$\xi_{ref} = 0$ Gas is not reformed $\xi_{ref} = 1$ Gas is completely reformed

$\xi_{ox} = 0$ Gas is not oxidised $\xi_{ox} = 1$ Gas is completely oxidised

Note that the extent of oxidation reaction is identical to the well known fuel utilisation, relating the rate of electrochemical fuel consumption to the fuel feed rate. Usually, a high fuel utilisation is desirable.

### 8.1.2
### Equations

Replacing the molar fractions by conversion numbers is achieved with the following equation:

$$\chi_i = \frac{\chi_{i,\,feed} + \sum_{j=ox,\,ref} v_{i,j} \xi_j \xi_{j,\,max}}{1 + \sum_{j=ox,\,ref} \bar{v}_j \xi_j \xi_{j,\,max}} \tag{8.2}$$

where the feed composition can be described by the steam-to-carbon ratio, $S/C$:

$$\chi_{CH_4, feed} = \frac{1}{1 + S/C} \tag{8.3}$$

$$\chi_{H_2O, feed} = \frac{S/C}{1 + S/C} \tag{8.4}$$

The constant $\xi_{j, max}$ describes the maximum stoichiometric conversion and are used to scale the conversions to unity. If only water and methane are used as fuel gases in a steam-to-carbon ratio, $S/C$, they read:

$$\xi_{ref, max} = \frac{1}{1 + S/C} \tag{8.5}$$

$$\xi_{ox, max} = \frac{4}{1 + S/C} \tag{8.6}$$

In addition to this, the possibility of anode gas recycling shall be considered. Figure 8.2 shows the assignments of all molar gas flows inside and outside the anode channel. At steady state, the flows differ due to two reasons: firstly the addition of the recycle flow to the anode inlet, and secondly due to the change of total mole numbers with reactions. To eliminate the influence of the second mentioned, a corrected molar flow is introduced, marked with a tilde sign:

$$\Gamma = \tilde{\Gamma} \cdot \left( 1 + \sum_{j = ref, ox} \bar{\nu}_j \cdot \xi_j \xi_{j, max} \right) \tag{8.7}$$

From the total and the partial molar balance at the mixing point before the channel inlet, the mixing rules can be derived:

$$\tilde{\Gamma}_{in} = \tilde{\Gamma}_0 + \tilde{\Gamma}_{back} \tag{8.8}$$

$$\xi_{j, in} = \frac{\tilde{\Gamma}_0 \cdot \xi_{j, 0} + \tilde{\Gamma}_{back} \cdot \xi_{j, back}}{\tilde{\Gamma}_0 + \tilde{\Gamma}_{back}} \tag{8.9}$$

Applying the listed assumptions to the anode mass balances from the reference model and inserting these definitions, one obtains the following ordinary differential equations, which are the main equations of the steady state anode model. For a more detailed derivation of these, we refer to [2]:

$$\frac{d\xi_j}{d\zeta} = \frac{Da_j r_j}{\tilde{\Gamma}} \cdot \frac{1}{\xi_{j, max}} \tag{8.10}$$

with

$$j = \{ref, ox\}$$

The boundary conditions are given at the inlet:

$$\xi_j(\zeta = 0) = \xi_{j,\text{in}} \tag{8.11}$$

This model can not only be used to simulate anode channels with direct internal reforming with different Damkoehler numbers for these processes. Setting $Da_{\text{ref}} = 0$ results in an anode channel without reforming reactions, and a reforming reactor can be obtained by $Da_{\text{ox}} = 0$.

In addition to the two ODEs, the reaction kinetics are required. The kinetic expressions are given here in a short and general form, without going into the details of the reaction orders. The reforming reaction rate depends on the gas composition and the given temperature parameter, the equilibrium constant is only temperature dependent:

$$r_{\text{ref}}(\xi, \vartheta) = k_{\text{ref}}(\vartheta) \cdot \left( r_{\text{ref}}^+(\xi) - \frac{r_{\text{ref}}^-(\xi)}{K_{\text{eq,ref}}(\vartheta)} \right) \tag{8.12}$$

The oxidation rate depends not only on the gas composition and the temperature parameter, but also on the electric potential difference at the charged double layer of the anode electrode, $\Delta\phi_a$. In addition, the oxidation rate at a certain location $\zeta$ must be identical to the cathodic reduction rate at that point. The reduction reaction depends on the electric potential difference at the cathode electrode, $\Delta\phi_c$. Describing both reactions by Butler–Volmer type reaction kinetics, we obtain:

$$r_{\text{ox}}(\xi, \vartheta, \Delta\phi_a) = k_{\text{ox}}(\vartheta, \Delta\phi_a) \cdot \left( r_{\text{ox}}^+(\xi) - \frac{r_{\text{ox}}^-(\xi)}{K_{\text{eq,ox}}(\vartheta, \Delta\phi_a)} \right) \tag{8.13}$$

$$r_{\text{red}}(\xi, \vartheta, \Delta\phi_c) = k_{\text{red}}(\vartheta, \Delta\phi_c) \cdot \left( r_{\text{red}}^+(\xi) - \frac{r_{\text{red}}^-(\xi)}{K_{\text{eq,red}}(\vartheta, \Delta\phi_c)} \right) \tag{8.14}$$

The two Eqs. (8.13) and (8.14) contain four unknowns, namely $r_{\text{ox}}$, $r_{\text{red}}$, $\Delta\phi_a$ and $\Delta\phi_c$. Consequently, two additional equations are required. One of them describes the equality of oxidation and reduction rate at each location $\zeta$:

$$Da_{\text{ox}} \cdot r_{\text{ox}} = Da_{\text{red}} \cdot (-r_{\text{red}}) \tag{8.15}$$

The second additional equations state that the sum of the potential differences at the anode and the cathode equals the given cell voltage:

$$\Delta\phi_a + (-\Delta\phi_c) = U_{\text{cell}} \tag{8.16}$$

Equations (8.13) to (8.16) have to be solved for the four unknowns, of which $r_{\text{ox}}$ is of interest in Eq. (8.10). To describe the electrical power output of the system, the product of cell voltage and oxidation conversion is used:

$$P_{el} = \frac{\Delta\xi_{ox} U_{cell}}{F} \tag{8.17}$$

where $\Delta\xi_{ox}$ is the fuel utilisation or the total electric current at dimensionless feeding rate equal unity.

To summarise this section, the relevant equations of the steady state anode models are specified. Outside any channels, Eqs. (8.8) and (8.9) describe mixing rules. These might be necessary to calculate the boundary condition (Eq. (8.11)) for Eq. (8.10), which together with the reaction rate expressions in Eqs. (8.12) to (8.13) describe the states inside such channels. To solve Eq. (8.13) for the required oxidation rate, $r_{ox}$, Eqs. (8.14) to (8.16) are needed. The concentration terms inside the reaction kinetic expressions, the same terms and reaction orders are used as in the kinetics of the reference model.

### 8.1.3
### Conversion Diagram

With only two locally distributed states in the model, it is possible to illustrate the results of the steady state anode model in a phase diagram. In this kind of plot, the independent spacial variable is eliminated by plotting $\xi_{ox}$ versus $\xi_{ref}$. Figure 8.3 shows the extent of the reforming reaction on the horizontal axis and the extent of oxidation reaction on the vertical axis. Because all molar fractions must be positive, the area of physically reasonable states is limited. The extents of reaction may only be situated within the triangle in Fig. 8.3.

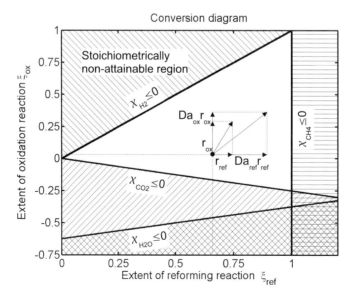

**Fig. 8.3** The conversion diagram, stoichiometrically attainable region and reaction rate vector.

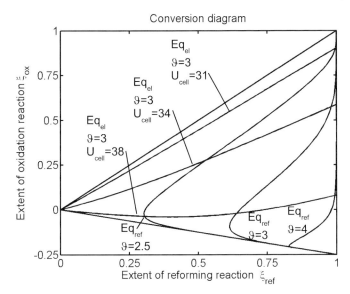

**Fig. 8.4** Equilibrium lines of the reforming reaction ($Eq_{ref}$) at various temperatures and equilibrium lines of the oxidation reaction ($Eq_{el}$) at constant temperature and various cell voltages.

For a given temperature and cell voltage, the reaction rates at every point within the diagram are defined. A positive reforming reaction rate means that the gas state moves to the right, while a positive oxidation rate causes a move upwards. Together, both reaction rates form a vector indicating in which direction the state moves along the spacial coordinate.

Because all reactions are considered reversible reactions, the gas composition may reach equilibrium for one or both reactions. In Fig. 8.4, several equilibrium lines are displayed for different temperatures and voltages. On these lines, the reaction rate of the corresponding reaction becomes zero. The reforming equilibrium lines start from the right bottom corner and end in the right upper corner of the admissible region. How far they reach to the left only depends on the given temperature. In general, low temperatures come with a low equilibrium conversion of the reforming reaction, thus the equilibrium curve reaches far to the left. Reforming equilibrium lines can be calculated from Eq. (8.12) by setting the rate to zero and then solving for a function $\xi_{ref}(\xi_{ox})$, at which the reforming reaction rate is zero.

The oxidation equilibrium lines always start at the diagrams origin point and end somewhere on the right boundary line. Their course depends on both temperature and cell voltage. The three exemplary lines in Fig. 8.4 are at the same temperature, but different cell voltage. On such a line, the oxidation process comes to a halt. The oxidation equilibrium line is calculated from Eqs. (8.13) to (8.16), where both electrochemical reaction rates are set to zero. The three result-

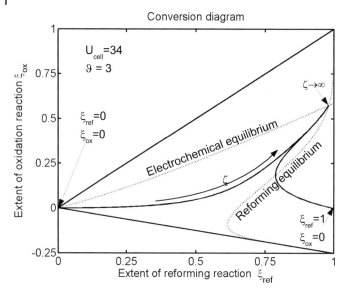

**Fig. 8.5** Conversion diagram with two exemplary trajectories for identical temperature and cell voltage, but with different inlet conditions.

ing equations then contain two independent variables, which are both extents of conversion, and two unknown potential differences. This allows to solve for a function $\xi_{ox}(\xi_{ref})$, at which the electrochemical equilibrium is located. Due to their characteristic course, there is always exactly one intersection point of both curves for a given temperature and cell voltage. At this point, both reaction rates become zero. Thus this is a stationary point. Because the reactions always run towards their equilibrium, this stationary point is an attractor in the phase diagram.

Two exemplary trajectories are shown in Fig. 8.5. The boundary condition of the first simulation is that both extents of reaction are zero, meaning that the gas is neither reformed or oxidised at the channel entrance. Upon entering, the reforming reaction starts and as soon as there is a significant amount of hydrogen available, the oxidation reaction rate also increases. Thus the trajectory moves from the origin to the right and then starts to move upward as well. For an infinite channel length, the trajectory ends in the attractor point. The same holds true for a different inlet condition. Assume a completely reformed gas entering the anode channel. It ends up in the same attractor point as does the first trajectory, but on its way, the reforming reaction runs in backward direction and the reforming equilibrium line is crossed in a vertical manner.

With this, the equations and the illustration tool of the steady state anode model are complete. In the following section, some applications of this tool are shown.

## 8.2
**Applications of the Steady State Anode Model**

The steady state anode model can be applied for various purposes. In the following, four applications are demonstrated. These are:
- comparison of different combinations of reforming concepts,
- evaluation of the benefit of a fuel cell cascade,
- evaluation of a fuel cell operated with an anode exhaust gas recycle,
- evaluation of a fuel cell with sidefeed in the anode channels.

### 8.2.1
**Comparison of Reforming Concepts**

As indicated in Fig. 8.1, three different reforming concepts are available for high temperature fuel cells. The steady state anode model presented above allows a comparison of various combinations of reforming concepts. First, a system without reforming catalyst inside the anode channel is considered, i.e. a fuel cell without DIR. Four alternatives for fuel gas treatment are discussed:
- Low-temperature ER which can easily be heated by the cell's hot exhaust gas.
- IIR where the reforming temperature equals that of the cell.
- High-temperature ER which requires additional external heating.
- Sidefeed of fresh fuel gas into the anode.

In this application, all three reforming units are considered to be infinitely long, thus equilibrium is reached in each unit. The effluent of the reforming unit is fed into the anode channel where only hydrogen oxidation takes place ($Da_{ref} = 0$). The resulting trajectories are plotted in Fig. 8.6. Low temperature ER and IIR produce anode fuel gases which are reformed to an extent of 30% and 65%, respectively. The subsequent fuel cell process can reach fuel utilisations of 20% and 60%, respectively. This fuel utilisation is not satisfying because a large portion of the fuel cannot be utilised for electric energy production. Only high-temperature ER offers the possibility to reach high fuel utilisation, but its drawback is the need for extra heating, which surely will decrease the overall system efficiency. Thus, none of these alternatives is satisfying.

The situation changes significantly when DIR is applied. Figure 8.7 shows that independent of the gas pre-treatment, high fuel utilisations can be reached with DIR, because all trajectories end in a single point being close to the right upper corner. Here, an additional drawback of the high-temperature ER is revealed: As the cell temperature is lower than the reforming temperature, the anode inlet condition is beyond the equilibrium line of the reforming reaction and thus the reforming runs to the backward direction in the first part of the channel, spoiling

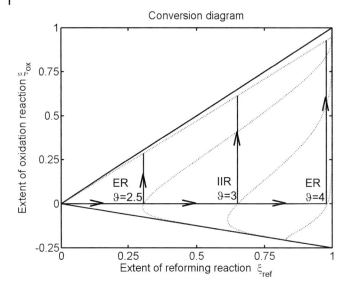

**Fig. 8.6** Conversion diagram with trajectories for different reforming configurations without DIR.

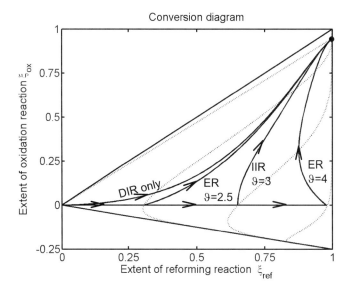

**Fig. 8.7** Conversion diagram with trajectories for different reforming configurations based on DIR. High degrees of fuel utilisation are attainable. Compare to Fig. 8.6.

the high conversion degree of reforming. The same attractor point, being reached by these three configurations with gas pre-treatment before the anode channel, can be reached by a system without any pre-treatment, where the fresh gas is directly fed into the anode channel (trajectory 'DIR only', Fig. 8.7).

These considerations clearly show the advantage of the DIR concept. Independent of the pre-treatment of the feed gas, the application of DIR always leads to a high degree of fuel utilisation. As shown here, the steady state anode model and the representation of its solutions in the conversion diagram are useful tools for evaluation and comparison of different process configurations.

### 8.2.2
### Fuel Cell Cascades

Various authors consider appropriate flow sheets for fuel cell systems. Fellows [3] proposes to employ a cascade of fuel cells, operating at different cell voltages, to increase the overall electric power output. According to this concept the exhaust gas of the first cell's anode compartment is fed into a second cell and so on (Fig. 8.8). The cascade considered here consists of two cells. Both cells together have the same size as the single cell used as a reference process for comparison.

Concerning the electric connections of the cascade, two different configurations are possible (Fig. 8.9). One possibility is that each cell is operated at its own cell voltage ($U_{cell, 1}$ and $U_{cell, 2}$) with an independent cell current ($I_{cell, 1}$ and $I_{cell, 2}$). In this case, both cells require a separate electrical converter and they are connected

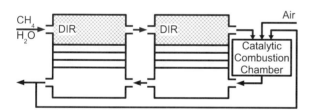

**Fig. 8.8** Flow sheet of a fuel cell cascade.

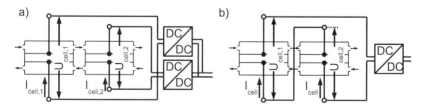

**Fig. 8.9** Alternative electric configurations of a fuel cell cascade:
(a) parallel cascade with independent cell voltage and cell current of each cell; (b) series cascade with independent cell voltage and equal cell current.

as parallel current sources (Fig. 8.9a). The alternative is a configuration where the individual cell voltages differ, while the same cell current runs through both cells (Fig. 8.9b). This series connection requires only one electrical converter, thus it needs less hardware installation. On the other hand, the equality condition of both cell currents reduces the number of degrees of freedom for the operation of this cell. In the following, we will restrict our analysis to the parallel configuration.

Three systems, that is the single cell, the two cell cascade and a twelve cell cascade are simulated with the steady state anode model. Their cell voltages are optimised to yield an optimum of electric power at a given feed rate. In any of the cascades, the size of each cell is identical and the sum of their sizes equals the size of the single cell system.

Figure 8.10 shows the resulting trajectories for the two cell cascade in parallel configuration in comparison to the single cell. As indicated by the striped area, the electrical power output from two cell cascade is 3.9% higher than the single cell power (see Table 8.1). Note that the activity of the electrodes and the reforming catalyst is equal in all cells and that both systems are fed with the same amount of fuel gas. Thus the electrical efficiency of the cascade system is higher than with the single cell, so a cascade could be an interesting design option. From an economical point of view, the additional hardware installiations which are necessary for such a system stand against this configuration, at least for small systems. For large systems, for example in the Megawatt class, this might be profitable. The benefit of the steady state anode model in this case goes beyond illustration and intuitive understanding, it can also serve as a tool for a first quantitative system design and it allows rough economic calculations.

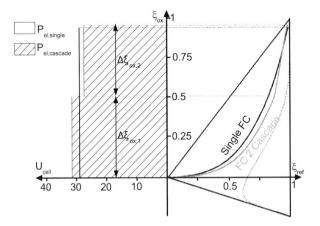

**Fig. 8.10** Trajectories for a single cell and two cell cascade in electric parallel configuration. Cell voltages are optimised for maximum electric power output. For quantitative results see Table 8.1.

**Table 8.1** Voltages, fuel utilisations and power output of a single cell, a two cell cascade and a twelve cell cascade.

|  | Voltage $U_{cell}$ | Fuel utilisation $\Delta\xi_{ox}$ | Electric power $P_{el}$ |
|---|---|---|---|
| Single cell | 28.8 | 0.947 | 62.4 |
| Cell 1 | 31.4 | 0.513 | 36.8 |
| Cell 2 | 27.3 | 0.449 | 28.0 |
| Cascade 2 | – | 0.962 | 64.8 (+3.9%) |
| Cascade 12 | – | 0.957 | 65.7 (+5.3%) |

Of course, a two cell cascade is not the only option for fuel cell cascades. Further splitting of the cell area leads to cascades with more than two cells, for example a cascade with twelve cells. Figure 8.11 shows the trajectory of such a twelve cell cascade with optimal voltage distribution. The overall electrical power is about 5.3% higher than the power from the single cell. This system is a good approximation of a virtual fuel cell system, where the cell voltage can be arbitrarily varied along the spacial coordinate. The trajectory in Fig. 8.11 is an approximation of the optimal trajectory in such a system.

This diagram is analogous to temperature plots in tube reactors in chemical engineering. In this classical field, temperature is a spacially distributed variable that needs to be controlled carefully in order to obtain maximum conversion and selectivity of a desired product. In electrochemical systems like fuel cells, the cell voltage takes this role of a control variable. There exists an optimal spacial distribution of the cell voltage at which the electrical power output becomes maximal.

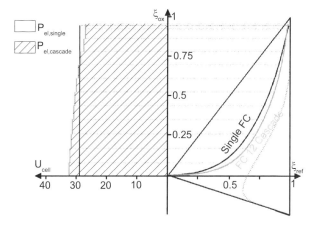

**Fig. 8.11** Trajectories for a single cell and twelve cell cascade. Cell voltages are optimised for maximum electric power output. For quantitative results see Table 8.1.

8.2.3
**Anode Exhaust Gas Recycling**

In the scientific literature one can also find the idea to recycle a part of the anode exhaust gas back towards the anode inlet [3], as illustrated in Fig. 8.2. Especially for pure DIR systems this has certain advantages: If fresh feed gas is fed into the anode channel, only small hydrogen contents will be present at the inlet and thus the electrode will not be used in this region. This can be amended by recycling a part of the exhaust, which still contains some hydrogen, back to the inlet. Define the recycle flow via a recycle ratio, $R$, according to:

$$R = \frac{\tilde{\Gamma}_{\text{feed}}}{\tilde{\Gamma}_{\text{back}}} \tag{8.18}$$

Then, the mixing rules (Eqs. (8.8) and (8.9)) simplify to

$$\tilde{\Gamma}_{\text{in}} = (1 + R) \cdot \tilde{\Gamma}_{\text{feed}} \tag{8.19}$$

$$\xi_{j,\text{in}} = \frac{\xi_{j,0} + R \cdot \xi_{j,\text{back}}}{1 + R} \tag{8.20}$$

where the initial condition is identical to Eq. (8.11):

$$\xi_j \left( \zeta = 0 \right) = \xi_{j,\text{in}} \tag{8.21}$$

and the extent of reaction of the recycle stream is equal to the conditions at the channel outlet:

$$\xi_{j,\text{back}} = \xi_j \left( \zeta = 1 \right) \tag{8.22}$$

Figure 8.12 shows the trajectory of a single cell without ($R = 0$) and with anode gas recycling ($R = 0.3$). Both are operated at their individual optimal cell voltage, so that the electric power output is maximised. While the inlet condition of the system without recycle is at the origin, the inlet condition of the recycle system is calculated via the lever rule (Eq. (8.20)) from the conditions of the feed gas and the anode exhaust gas. Because the exhaust gas composition is unknown, first one has to guess it, then calculate the inlet condition and integrate along the channel to obtain a new value for the outlet condition. For the correct solution, the guessed value and the simulated one have to be equal. This can be achieved by an error minimisation algorithm or by iterative insertion of the latest simulated value as a new guess for the exhaust condition. The electric power output of the recycle system is always lower than what one gets from the system without recycling. This is due to the fact that back mixing always deteriorates the reactor performance in systems with positive reaction order with respect to the educts. Thus, the higher the recycling ratio is, the lower the electric power

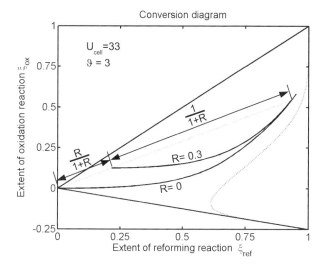

**Fig. 8.12** Conversion diagram with trajectories for systems without and with anode gas recycling.

output becomes. Moreover, the conversion diagram can be used for explaining the role of anode gas recycling. For DIR systems with unreformed feed, the trajectories are usually bent to the left-hand side. This means that the inlet condition of a recycle system can only lie within the bent of the corresponding no-recycle system. Consequently, the trajectory lies closer to the main diagonal. In this region, lower cell voltages must be applied to achieve comparable oxidation rates and therefore reach high fuel utilisations. Thereby the electric power output of the cell is decreased. This shows that back mixing has a negative effect on the fuel cell performance, a rule that is well known in chemical engineering. Although this result shows the disadvantage of anode gas recycling, this concept might still have some benefits that compensate for the loss in power output. This is due to the fact that feeding fresh gas into a DIR channel without any gas recycling leads to high reforming reaction rates at the inlet. Consequently, the temperature drops sharply leading to strong temperature gradients. This can be avoided by recycling of anode exhaust gas which leads to a homogenisation of the temperature profile.

### 8.2.4
### Fuel Gas Sidefeed

Another concept which is well-known in chemical engineering is the sidefeed in tube reactors. The idea is to feed fresh educts into a long reactor in order to manipulate concentration levels inside the reactor. This concept can be carried over to high temperature fuel cells in such a way that fresh fuel gas is fed into the anode channel at discrete locations (see Fig. 8.13).

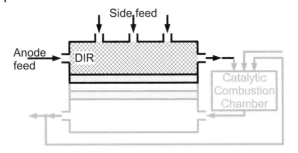

**Fig. 8.13** MCFC with sidefeed in the anode channel.

Such a side feed fuel cell can be analysed with the steady state anode model. Figure 8.14 shows the trajectory of such a system with three side feeds. Each segment between two side feeds has the same length, and the cell voltage is constant throughout the whole cell. For easier comparison, the nearly optimal trajectory of the twelve cell cascade is also displayed here. As can be seen from this figure, a side feed always interrupts the trajectory and draws it towards the origin, which is the state of the feed gas. In this figure, the side feed is used to manipulate the trajectory in such way that it approximates the optimal one. Nevertheless, when the side feed parameters and the cell voltage are submitted to an optimisation, the optimum has no side feed at all and the cell voltage is identical to the optimum for a single cell (Section 8.8).

While the idea of approximating the optimal trajectory is good, the side feed concept fails for two reasons:

- A side feed is always a sort of back mixing. As shown in Section 8.2, this is disadvantageous with regard to the cell power.

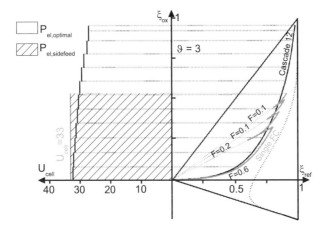

**Fig. 8.14** Trajectory of a system with side feed.

- Besides the thermodynamic equilibrium, the maximum fuel utilisation is further limited in side feed reactors. If the starting point and the end point of a part of the trajectory are both lying on a straight line through the origin, no higher fuel utilisation can be obtained (see last part of the trajectory in Fig. 8.14)

Consequently, any side feed has to be avoided in MCFC. In analogy from chemical reactor engineering, this holds as long as the reactions show a positive reaction order with respect to the educts.

## 8.3
## Summary

The MCFC offers the possibility of an integration of the endothermic reforming process and the exothermic electrochemical oxidation process inside one single reactor. In the direct internal reforming concept, also mass integration of both processes helps to overcome equilibrium limitations of the reforming reactions. Thereby, process integration allows for high degrees of fuel utilisation and high efficiency of fuel cell systems.

The steady state anode model focuses on the description of these two processes and their interplay. It is a manageable isothermal model, based on balances of mass and charge. The model describes the extent of the reforming and the oxidation reactions along the anode channel. The essential simulation results can be easily displayed in a conversion diagram which is a phase diagram of the two dynamic state variables, namely the extents of the two reactions.

Within this chapter, four applications of this tool are demonstrated. The comparison of different reforming concepts like direct and indirect internal reforming as well as external reforming reveals that the direct integration of both processes is advantageous. In the second application, the model is used to evaluate the benefit of a fuel cell cascade with individual cell voltages for each cell. It turns out that cell cascades can significantly improve the overall system performance. Cell voltage is an important process control variable in such systems, and an optimal spacial voltage exists, which can be approximated in a fuel cell cascade. In the third application we demonstrate that recycling of anode exhaust gas always leads to a decrease in electrical power output. As a last design option, the sidefeed of fresh fuel gas into the anode is considered. Because this is also a kind of backmixing like in the exhaust recycling option, it always decreases the overall cell performance.

This chapter shows that the steady state anode model is well suited for the discussion of conceptual design issues of the integrated reforming concept. Due to the abstract formulation with conversions instead of concentrations as state variables, this model can be directly applied to other high temperature fuel cells like SOFC, too.

## Bibliography

1 Heidebrecht, P., Modelling, Analysis and Optimisation of a Molten Carbonate Fuel Cell with Direct Internal Reforming (DIR-MCFC), VDI Fortschritt-Berichte, Reihe 3, Nr. 826, VDI-Verlag, Düsseldorf, 2005.

2 Heidebrecht, P., Sundmacher, K., Conceptual design of the integration of the reforming process in high temperature fuel cells, Journal of Power Sources 145, 1, 2005, 40–49.

3 Fellows, R., A novel configuration for direct internal reforming stacks, Journal of Power Sources 71, 1998, 281–287.

# Part III
# Optimization and Advanced Control

*Molten Carbonate Fuel Cells.* Edited by Kai Sundmacher, Achim Kienle, Hans Josef Pesch,
Joachim F. Berndt, and Gerhard Huppmann
Copyright © 2007 WILEY-VCH Verlag GmbH & Co. KGaA, Weinheim
ISBN: 978-3-527-31474-4

# 9
# Model Reduction and State Estimation

*Markus Grötsch, Michael Mangold, Min Sheng, and Achim Kienle*

## 9.1
## Introduction

The process operation of industrial high temperature fuel cell stacks like the Hot-Module is made difficult by the limited amount of measurement information available on-line. In the case of the HotModule, gas temperature measurements can be taken at the outlet of the anode and of the cathode gas channels. A gas chromatograph provides information about the gas composition at the channel outlet, however with large gaps of about one hour between two measurements. Measurement of temperatures inside a molten carbonate fuel cell stack is difficult due to the aggressive nature of the salt melt that corrodes thermocouples rapidly. Concentration measurements inside a stack are nearly impossible. For manual operation of the HotModule as well as for automatic control, additional information on the internal state of the stack would be highly desirable. Without such information, large safety factors, e.g. concerning the rate of change of the cell current, have to be applied during process operation in order to prevent overheating, coking, or other damages. Thus, the small number of on-line measurements limits the maximum achievable current and power density and prevents one from tapping the full potential of the system.

The problems caused by a lack of on-line measurement information may be overcome by using a state estimator or observer (see e.g. [9]). The principle of an observer is illustrated in Fig. 9.1. An observer consists of a process model and a correction term. The process model is provided with all known inputs to the system and is used to simulate the process. If the simulation matched reality perfectly, one could use the simulated states directly as an estimate of the non-measurable states of the real system. In reality, however, the simulations will always deviate from the reference system because of uncertainties in the initial conditions, unknown disturbances, and inevitable model errors. In order to compensate these deviations, a correction term is added to the process model. The correction term depends on the difference between simulated and taken measurements. It has to be designed in such a way that the complete estimated state

*Molten Carbonate Fuel Cells.* Edited by Kai Sundmacher, Achim Kienle, Hans Josef Pesch,
Joachim F. Berndt, and Gerhard Huppmann
Copyright © 2007 WILEY-VCH Verlag GmbH & Co. KGaA, Weinheim
ISBN: 978-3-527-31474-4

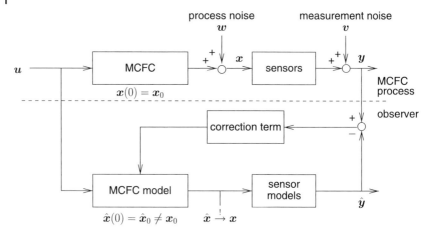

**Fig. 9.1** Principle of an observer.

vector converges towards the real state vector in finite time. This is only possible, if sufficient measurement information is available, i.e. if the system is observable. Consequently, the development of an observer for the HotModule requires three main steps: (1) the choice of a suitable process model, (2) the choice of a sensor combination that makes the system observable, and (3) the design of an observer correction. These steps will be discussed in the following.

## 9.2
### Development of a Nonlinear Reduced Model

In Chapter 3 of this book, a detailed spatially two-dimensional model of the Hot-Module is presented. The results of Chapter 5 show that this model is able to describe the steady state and dynamic behaviour of the HotModule accurately and qualify it e.g. for the application to design problems. But for the state estimation problem considered here, the spatially distributed reference model is less suitable: its high complexity makes the numerical solution rather slow and expensive; furthermore the design of a state estimator on the basis of such a model is difficult. Therefore, a reduced model of low order is desirable. The reduced model should approximate the reference model's behaviour for the relevant operation conditions with reasonable accuracy. Its numerical solution should be considerably simpler than that of the reference model, and the design of an observer correction term should be easier. As the reduced model should be applicable to a large range of operation conditions, a linear model reduction is not sufficient. Instead, a nonlinear reduced model is developed in the following.

The model reduction is done in two steps. A first slight simplification of the detailed model is achieved by reducing the number of dynamic degrees of freedom. The time constants of the electric potential equations and the time constants in the gas phases are much smaller than the time constant of the energy balance of the solid. As the slow dynamics of the solid temperature dominates many process control problems, the electric potential fields as well as the anode gas channels can be considered as quasi-stationary. However, the use of a dynamic equation for the cathode gas temperature turns out to be advantageous for the numerical solution. The reason is that a change of the cell current causes a jump of the cathode gas temperature if quasi-stationarity is assumed.

A much stronger reduction of the system order is possible if the spatially distributed model can be approximated by a lumped system. For such a model reduction of parabolic partial differential equations, orthogonal projection methods have become a frequently used technique [1, 6]. One of these methods, the Karhunen–Loève-Galerkin procedure, is used in this work. The procedure is illustrated here for the example of the solid temperature equation that can be written in the form

$$c_p^S \frac{\partial \vartheta^S}{\partial \tau}\bigg|_{\tau,\zeta_1,\zeta_2} = \frac{1}{Pe_1^S} \frac{\partial^2 \vartheta^S}{\partial \zeta_1^2}\bigg|_{\tau,\zeta_1,\zeta_2} + \frac{1}{Pe_2^S} \frac{\partial^2 \vartheta^S}{\partial \zeta_2^2}\bigg|_{\tau,\zeta_1,\zeta_2} + \sigma_\vartheta^S \tag{9.1}$$

$(0 < \zeta_1 < l_1, 0 < \zeta_2 < l_2)$ with a nonlinear source term $\sigma_\vartheta^S$, and boundary conditions

$$\frac{\partial \vartheta^S}{\partial \zeta_1}\bigg|_{\tau,0,\zeta_2} = \frac{\partial \vartheta^S}{\partial \zeta_1}\bigg|_{\tau,l_1,\zeta_2} = \frac{\partial \vartheta^S}{\partial \zeta_2}\bigg|_{\tau,\zeta_1,0} = \frac{\partial \vartheta^S}{\partial \zeta_2}\bigg|_{\tau,\zeta_1,l_2} = 0 \tag{9.2}$$

The basic idea of the projection method is to approximate the unknown variable $\vartheta^S$ by a finite sum of products $\tilde{\vartheta}^S$:

$$\tilde{\vartheta}^S(\zeta_1, \zeta_2, \tau) = \sum_{i=1}^{N} \vartheta_i^S(\tau) \varphi_i(\zeta_1, \zeta_2) \tag{9.3}$$

The functions $\varphi_i$ in (9.3) are space dependent orthogonal basis functions, which are chosen beforehand. The $\vartheta_i^S$ are time dependent amplitude functions that follow from Eq. (9.1). The approximative solution $\tilde{\vartheta}^S$ will not solve Eq. (9.1) exactly, but a nonzero residual

$$\text{Res} := c_p^S \frac{\partial \tilde{\vartheta}^S}{\partial \tau} - \frac{1}{Pe_1^S} \frac{\partial^2 \tilde{\vartheta}^S}{\partial \zeta_1^2} - \frac{1}{Pe_2^S} \frac{\partial^2 \tilde{\vartheta}^S}{\partial \zeta_2^2} - \sigma_\vartheta^S \tag{9.4}$$

will remain. Galerkin's method of weighted residuals [3] requires that this residual must vanish if weighted by a basis function, i.e.

$$\iint_\Omega \text{Res} \cdot \varphi_i \, d\zeta_1 \, d\zeta_2 \overset{!}{=} 0, \quad i = 1, \dots, N \tag{9.5}$$

where $\Omega$ is the space domain of the system. Inserting (9.3) into (9.5) leads to $N$ ordinary differential equations for the time dependent amplitude functions $\vartheta_i^S(\tau)$. The outlined procedure raises a couple of questions that need further discussion: (1) the choice of suitable basis functions $\varphi_i$, and (2) the incorporation of the boundary conditions (9.2).

### 9.2.1
### Choice of Basis Functions

The accuracy of the reduced model, i.e. its deviation from the detailed model, mainly depends on two factors. The first one is the number of terms considered in the series approximations. The second is the choice of the basis functions. Obviously, the best approximation with a low order model is achievable, if basis functions are used that are tailored to the specific problem. The Karhunen–Loève (KL) decomposition method allows to compute such problem-specific basis functions numerically [8, 13]. The KL decomposition method uses numerical simulation results at discrete time points, so-called snapshots, obtained from dynamic simulations with the detailed model. For the sake of simplicity, the method is illustrated here for the case of a spatially discretised system, but can be applied to a spatially continuous system in the same way. In the case of a spatially discretised system, the snapshots $v_j \in \mathbb{R}^n$, $j = 1, \ldots, J$ give the value of the state vector at discrete time points $t_j$, e.g. the values of the solid temperature on a grid of $n$ grid points. The task is to find vectors $\varphi_i$ that approximate the snapshots $v_j$ as well as possible in the sense that

$$\frac{\sum_{j=1}^{J} (\varphi_i, v_j)^2}{(\varphi_i, \varphi_i)} \overset{!}{=} \max, \tag{9.6}$$

where $(\cdot, \cdot)$ denotes the scalar product (see Fig. 9.2). The vectors $\varphi_i$ contain the values of the basis functions $\varphi(\zeta_1, \zeta_2)$ at the $n$ grid points. Equation (9.6) can be rewritten as

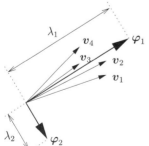

Fig. 9.2 Illustration of the Karhunen–Loève decomposition method.

$$\frac{(R\varphi_i, \varphi_i)}{\varphi_i, \varphi_i} \overset{!}{=} \max =: \lambda \tag{9.7}$$

with a symmetric $n \times n$ matrix $R$ given as

$$R = \frac{1}{J} \sum_{j=1}^{J} v_j v_j^T \tag{9.8}$$

The solution of Eq. (9.7) is the eigenvector $\varphi_1$ belonging to the largest eigenvalue $\lambda_1$ of $R$. The set of eigenvectors $\varphi_1, \varphi_2, \varphi_3, \ldots$ of $R$ to the eigenvalues $\lambda_1 > \lambda_2 > \lambda_3 > \cdots$ forms a set of orthogonal basis vectors, where $\varphi_1$ is the most typical structure of the snaphots, $\varphi_2$ is second, and so on. Consequently, the leading eigenvectors of $R$ can be seen as the discretised versions of the sought-after basis functions $\varphi_i(\zeta_1, \zeta_2)$.

Especially for large systems, the system order $n$ is often much larger than the number of snapshots $J$. In such a case, it is advantageous to express $\varphi_i$ as a linear combination of the snapshots:

$$\varphi_i = \sum_{j=1}^{J} \alpha_j^i \cdot v_j \tag{9.9}$$

Then it can be shown [13] that Eq. (9.7) can be simplified to the eigenvalue problem

$$C a_i = \lambda_i a_i \tag{9.10}$$

where $a_i = (\alpha_1^i, \alpha_2^i, \ldots, \alpha_J^i)^T$ and $C$ is a $J \times J$ symmetric positive definite matrix with the following elements:

$$C_{kl} = \frac{1}{J}(v_k, v_l), \quad k, l = 1, \ldots, J \tag{9.11}$$

In summary, the construction of basis functions by KL decomposition requires the solution of the eigenvalue problem (9.10). The advantage of the method compared to other orthogonalisation methods like the Gram–Schmidt process is that the eigenvalues $\lambda_i$ can be interpreted as a measure of how well a basis function is able to approximate the average of the snapshots. If an eigenvalue is very small, this means that the corresponding basis function does not have much in common with the snapshots. Such an eigenfunction can be neglected in the approximation (9.3) without much loss of accuracy. Therefore, the eigenvalues can be used to determine the order $N$ of the reduced model that is necessary for a good approximation of the detailed model, at least for the scenarios under which the snapshots were obtained.

A somewhat open question is, how the snapshots should be chosen in order to get a complete picture of the possible solution profiles of the detailed model. Patera and co-workers developed guidelines for certain linear problems (see e.g. [14]), but to our knowledge there are no strict rules on how to choose snapshots for a nonlinear problem as considered here. A pragmatic approach is to use a high number of test simulations under different operation conditions. This is easily possible as the solution of Eq. (9.10) is rather cheap even for large values of $J$.

## 9.2.2
## Treatment of Boundary Conditions

There are several alternatives to take care of the boundary conditions within the Galerkin approach. The first possibility used e.g. in [13] is to include the boundary conditions (9.2) in the weak problem formulation (9.5) through integration by parts.

$$\iint_\Omega \left( c_P^S \frac{\partial \vartheta^S}{\partial \tau} - \frac{1}{Pe_1^S} \frac{\partial^2 \vartheta^S}{\partial \zeta_1^2} - \frac{1}{Pe_2^S} \frac{\partial^2 \vartheta^S}{\partial \zeta_2^2} - \sigma_\vartheta^S \right) \varphi_i \, d\zeta_1 \, d\zeta_2 = 0$$

$$\Leftrightarrow \iint_\Omega \left( c_P^S \frac{\partial \vartheta^S}{\partial \tau} - \sigma_\vartheta^S \right) \varphi_i \, d\zeta_1 \, d\zeta_2$$

$$- \frac{1}{Pe_1^S} \int_{\zeta_2=0}^{l_2} \left[ \frac{\partial \vartheta^S}{\partial \zeta_1} \right]_{\zeta_1=0}^{\zeta_1=l_1} \varphi_i \, d\zeta_2 - \frac{1}{Pe_2^S} \int_{\zeta_1=0}^{l_1} \left[ \frac{\partial \vartheta^S}{\partial \zeta_2} \right]_{\zeta_2=0}^{\zeta_2=l_2} \varphi_i \, d\zeta_1 = 0$$

$$\Leftrightarrow \iint_\Omega \left( c_P^S \frac{\partial \vartheta^S}{\partial \tau} - \sigma_\vartheta^S \right) \varphi_i \, d\zeta_1 \, d\zeta_2 = 0 \qquad (9.12)$$

The modified residual condition (9.12) can then be used to determine equations for the amplitude functions $\vartheta_i^S(\tau)$. However, using the modified residual condition cannot guarantee that the approximate solution resulting from the Galerkin method fulfills the boundary conditions accurately. In the case of the HotModule, this approach led only to poor results.

Another method suggested by Finlayson [3] for one-dimensional problems is to replace some of the residual conditions (9.5) by the boundary conditions (9.2). By this way, the approximate solution resulting from the Galerkin method will fulfill the boundary conditions exactly. However, there is no guideline, which residual conditions should be dropped, and the extension of the method to a two-dimensional problem is difficult.

Therefore, a different approach is tried in this work. Instead of dropping residual conditions, additional variables are introduced in order to fulfill the boundary conditions. The additional variables are the values of the states at the boundary points, e.g. $\vartheta^{S,*}(\zeta_2, \tau)$ for $\vartheta(0, \zeta_2, \tau)$. These variables are still undefined because of the way how the basis functions $\varphi_i$ are determined: The $\varphi_i$ follow from a numerical solution of the detailed reference model using a spatial discretisation on finite volumes, and hence are only known at discrete gridpoints $0.5 \cdot \Delta\zeta_{1/2}, 1.5 \cdot \Delta\zeta_{1/2}, \dots$ In the next step, the spatial gradient $\partial\vartheta^S/\partial\zeta_{1/2}$ is approximated by finite differences and inserted into the boundary conditions (9.2) in order to obtain equations for the newly introduced variables, e.g.

$$\frac{\partial \vartheta^S}{\partial \zeta_1}\bigg|_{0,\zeta_2} \approx \frac{\tilde{\vartheta}^S(0.5 \cdot \Delta\zeta_1, \zeta_2, \tau) - \vartheta^{S,*}(\zeta_2, \tau)}{0.5 \cdot \Delta\zeta_1} \overset{!}{=} 0 \tag{9.13}$$

Finally, the boundary values are used in the numerical quadrature of the integrals terms in Eq. (9.5). As Eq. (9.13) can be solved explicitly for $\vartheta^{S,*}$, the introduction of additional unknowns does not increase the order of the reduced model.

### 9.2.3
### Resulting Reduced Model of the HotModule

In order to obtain a reduced model of the MCFC system, not only the solid temperature profile has to be approximated by a set of basis functions, but also the space profiles of the gas temperatures, the total molar flow rates, the gas compositions, and the electric potential fields. The resulting reduced model is a differential algebraic system with a differential index of one that has the following structure:

$$\frac{d\vartheta^S}{dt} = f^S(\vartheta^S, \vartheta^A, \vartheta^C, x_i, \gamma^A, \gamma^C, \Delta\phi^A, \Delta\phi^C) \tag{9.14}$$

$$\frac{d\vartheta^C}{dt} = f^C(\vartheta^S, \vartheta^A, \vartheta^C, x_i, \gamma^A, \gamma^C, \Delta\phi^A, \Delta\phi^C) \tag{9.15}$$

$$0 = g(\vartheta^S, \vartheta^A, \vartheta^C, x_i, \gamma^A, \gamma^C, \Delta\phi^A, \Delta\phi^C) \tag{9.16}$$

In (9.14)–(9.16), $\vartheta^S = (\vartheta_1^S, \vartheta_2^S, \ldots, \vartheta_N^S)$ is the vector of the amplitude functions for the solid temperature, $\vartheta^A$, $\vartheta^C$, $x_i$, $\gamma^A$, $\gamma^C$, $\Delta\phi^A$, $\Delta\phi^C$ are the corresponding vectors of the time dependent amplitude functions for the anode and cathode gas temperatures, the gas compositions, the anode and cathode molar flux densities, and the electrical potential fields at the anode and at the cathode, respectively.

In order to determine suitable basis functions for the MCFC model, the response of the detailed model to an increase of the cell current and to a subsequent decrease to the original value is computed numerically. The necessary number of basis functions for each variable is found by comparing the simulation results of the reduced model with the results of the detailed model for this test simulation. Good results are obtained when using 8 basis functions for the temperatures, 10 basis functions for the total anode flow rate and the anode potential, 9 basis functions for cathode flow rate and the cathode potential, and between 2 and 10 basis functions for the molar fractions of the gas components. The resulting reduced model consists of 16 ordinary differential equations and 115 algebraic equations, thereby reducing the dynamic order of the system by a factor of more than 100 compared to the detailed model. Due to nonlinearities of the model, the evaluation of the right-hand sides of (9.14)–(9.16) requires a numerical quadrature. In spite of this fact, the numerical solution of the reduced model is two orders of magnitude faster than that of the detailed model. Test simulations with randomly varying cell current show that the deviations of the reduced model from the complete model are in almost all cases below 1% [10].

Although the basis functions $\varphi_i$ are computed numerically, it should be noted that the reduced model still depends explicitly on the physical and operation parameters of the detailed model, i.e. a change of a physical parameter value directly affects the behaviour of the reduced model. Simulations show that the extrapolation qualities of the reduced model with respect to various model parameters are quite good [11].

From these tests it is concluded that the reduced model describes the behaviour of the MCFC system with reasonable accuracy and can be used as a basis for a state estimator.

## 9.3
## Investigation of Observability

In the HotModule, the following on-line measurement information is available: Apart from electrical quantities like the total cell voltage, various temperature and concentration measurements are made. The temperatures of the gases at the outlet of several anode gas and cathode gas channels are measured. As the gases are mixed strongly at the outlet of the gas channels, these temperature measurements can only give information on a middle outlet temperature averaged over all anode gas channels and all cathode gas channels, respectively. The spatially averaged composition of the gases at the outlet of the anode and cathode gas channels is measured by a gas chromatograph with sampling times of about 1 h.

The observability of the system is investigated based on the reduced model. With the following definitions:

$$x_D := (\vartheta^{S\,T}, \vartheta^{C\,T})^T \tag{9.17}$$

$$x_A := (\vartheta^{A\,T}, x_i^{\,T}, \gamma^{A\,T}, \gamma^{C\,T}, \Delta\phi^{A\,T}, \Delta\phi^{C\,T})^T \tag{9.18}$$

the reduced model (9.14)–(9.16) can be written as:

$$\dot{x}_D = f(x_D, x_A) \tag{9.19}$$

$$0 = g(x_D, x_A) \tag{9.20}$$

Because the cell voltage depends nonlinearly on the state vector, a nonlinear output equation results for the measurement vector $y$:

$$y = h(x_D, x_A) \tag{9.21}$$

The observability analysis is based on the observability mapping, which consists of the output equation (9.21) and its time derivatives:

$$
\begin{pmatrix} \boldsymbol{y} \\ \dot{\boldsymbol{y}} \\ \vdots \\ \boldsymbol{y}^{(m-1)} \end{pmatrix} = \begin{pmatrix} \boldsymbol{h}(\boldsymbol{x}_D, \boldsymbol{x}_A) \\ \dfrac{\partial \boldsymbol{h}}{\partial \boldsymbol{x}_D} \dot{\boldsymbol{x}}_D + \dfrac{\partial \boldsymbol{h}}{\partial \boldsymbol{x}_A} \dot{\boldsymbol{x}}_A \\ \vdots \\ \cdots \end{pmatrix}
\tag{9.22}
$$

The time derivatives on the right-hand side of Eq. (9.22) can be eliminated with the help of the model equations (9.19) and (9.20). In order to prove global observability of a nonlinear system, one has to show that the observability mapping (9.22) can be solved uniquely for $\boldsymbol{x}_D$ and $\boldsymbol{x}_A$ [2, 15]. Due to the complexity of the reduced model, such a strict proof of global observability is hardly possible. Instead, the observability in the vicinity of a stationary operation point is checked here based on the linearised reduced model. The case of a measurable cell voltage and measurable outlet temperatures of the gases is considered, i.e. the measurement information from the gas chromatograph is not used. The Kalman criterion and the Hautus criterion are applied. Details can be found in [10], where also other sensor configurations are studied.

The Kalman criterion indicates a full rank of the observability matrix and hence full observability. By using the Hautus criterion, it is found that the modes closely connected to the amplitude functions of the cathode gas temperatures are well observable, whereas the observability of the modes connected to the amplitude functions of the solid temperature is worse. Especially the higher amplitude functions of the solid and cathode gas temperature are only weakly observable. However, it turns out that a poor estimate of the higher amplitude functions only has a minor effect on the estimate of the actual temperature profiles and the other physical states of the system.

## 9.4
## Design of an Extended Kalman Filter

As a state estimator for the HotModule, an extended Kalman filter with time discrete measurements and a time discrete correction term has been designed. In classical textbooks, e.g. [4], the extended Kalman filter is derived for systems of ordinary differential equations. However, an extension to differential algebraic systems with a differential index of one is straight forward [12]. The filter equations are briefly summarised in the following.

The actions of the filter can be divided into two steps: In a prediction step, a time continuous simulation of the reduced model is used to estimate the states of the system. In a time discrete update step that is performed at measurement times the model states are corrected. The corrected states are used as initial conditions for the subsequent prediction step. The two steps are described in more detail in the following.

During the prediction step, the model equations (9.19) and (9.20) and the equations for the error covariance matrix $P$

$$\dot{P} = FP + PF^T + Q \tag{9.23}$$

have to be solved. In (9.23), the Jacobian $F$ is the total derivative of the right-hand side vector of (9.19) with respect to $x_D$ after elimination of the algebraic states $x_A$:

$$F = \frac{\partial f}{\partial \hat{x}_D} + \frac{\partial f}{\partial \hat{x}_A} \frac{d\hat{x}_A}{d\hat{x}_D} \tag{9.24}$$

(The estimated filter states are denoted by a hat). The matrix $Q$ is a symmetric positive definite matrix, which should be equal the spectral density matrix of the process noise in the case of a linear system.

In the update step, the difference between the predicted measurement values and the actually taken measurement values is weighted by a gain matrix $K$ given by the following correlation:

$$K[HP(-)H^T + R] = P(-)H^T \tag{9.25}$$

where $P(-)$ is the predicted value of the error covariance matrix at the time of measurement; $R$ is the symmetric and positive definite covariance matrix of the measurement noise; $H$ is the Jacobian of the output equation (9.21) after elimination of $x_A$:

$$H = \frac{\partial h}{\partial \hat{x}_D} + \frac{\partial h}{\partial \hat{x}_A} \frac{d\hat{x}_A}{d\hat{x}_D} \tag{9.26}$$

The weighted difference between predicted output values $\hat{y}$ and measured output values $y$ is used to update the dynamic states of the system according to:

$$\hat{x}_D(+) = \hat{x}_D(-) + K(y - \hat{y}) \tag{9.27}$$

In (9.27), $\hat{x}_D(-)$ is the predicted state vector at the measurement time, $\hat{x}_D(+)$ is the corrected state vector used as an initial condition for the subsequent prediction step.

Finally, the update step corrects the error covariance matrix according to:

$$P(+) = [I - KH]P(-) \tag{9.28}$$

The gradient $d\hat{x}_A/d\hat{x}_D$ required in (9.24) and (9.26) is obtained by deriving (9.20) with respect to $\hat{x}_D$:

$$0 = \frac{\partial g}{\partial \hat{x}_D} + \frac{\partial g}{\partial \hat{x}_A} \frac{d\hat{x}_A}{d\hat{x}_D} \tag{9.29}$$

Because the reduced model has a differential index of one, (9.29) is regular and can always be solved for $d\hat{x}_A/d\hat{x}_D$.

For the design of the filter, the matrices $Q$ and $R$ are used as design parameters. $Q$ and $R$ are assumed to be diagonal matrices; the elements on the main diagonal are chosen heuristically by simulations. In the implementation of the Kalman filter for the MCFC, the Jacobians $F$ and $H$ are computed numerically by finite differences. Because the measurement intervals are considerably smaller than the dominating time constants of the system, the variation of $F$ is rather small between two measurement times. In order to increase the numerical efficiency, $F$ is computed in every update step and is kept constant in the prediction steps.

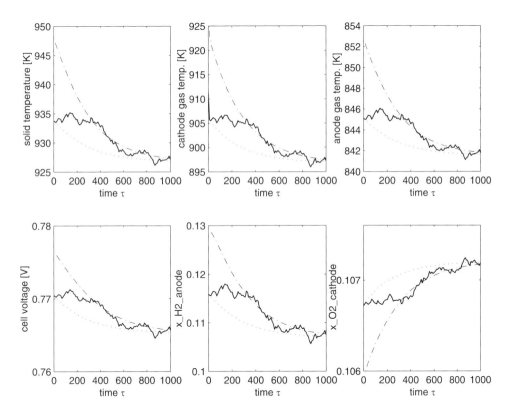

**Fig. 9.3** Convergence of the extended Kalman filter, if a transition of the real system is used as a reference (dash-dotted lines); transitions of the filter states without a correction at measurement times (dotted lines) and with a correction at measurement times (solid lines); first row from left to right: estimated solid temperature, estimated cathode gas temperature and the estimated anode gas temperature; second row from left to right: the estimated cell voltage, the $H_2$ molar fraction in the anode gas, the $O_2$ molar fraction in the cathode gas. The figure shows the temperatures in the middle of the electrode plane of the MCFC.

## 9.5
## Simulation Results

In a first step, the filter is tested with simulated measurement values generated by the reduced model perturbed with Gaussian noise. Standard deviations of 5 K for the average anode gas temperature, 5 K for the average cathode gas temperature and also 5 mV for the cell voltage are used. The size of the noise corresponds to the accuracy of real measurement values from the HotModule. The initial condition for each of the filters is a steady state for a different value of the cell current. At time 0, the cell current in the filter model is set to the correct value. A measurement interval of 10 time units (about 60 s) is assumed.

Figure 9.3 shows the results of the extended Kalman filter when a transition of the real system is used as reference. At time 0, the real system is at a steady state for a (dimensionless) cell current of 0.65. The extended Kalman filter is at steady state for a cell current of 0.68. At the beginning of the simulation, the cell current is increased to 0.7. It is found that after a short transient, the filter tracks the dynamic reference state correctly. The convergence is clearly faster than the transient of the uncorrected model to the steady state.

## 9.6
## Experimental Results

The extended Kalman filter is tested at the HotModule in Magdeburg. The filter is implemented in ProMot [17] and is simulated with DIVA [7] on a laptop computer (1700 MHz Pentium M processor). The laptop computer is connected to the process control system of the HotModule via a local area network. The filter is fed with input and measurement data read periodically from the process control system. The validation of the filter is done by using additional concentration measurements that are not used for the filter correction. The concentration measurements are obtained from a gas chromatograph at the outlet of the anode gas channels in time intervals of about two minutes.

Figure 9.4 depicts the estimation result for a steady state of the HotModule. The filter is started with wrong initial values but shows good convergence behaviour towards the real measurement variables with only a small stationary error. The filter convergence towards the concentration measurements is quick and of acceptable accuracy. The remaining small deviations are in the order of magnitude of direct concentration measurements. They can be explained by inevitable model errors. Tests with steady states at other operating conditions show comparable results. The estimates, especially the cell voltage, are perturbed by measurement noise of the input flow rate of the anode gas channel and by measurement noise of the cell current. Both quantities are input variables and its noise cannot be suppressed by the Kalman filter.

Figure 9.5 shows the estimation results if the HotModule undergoes two load changes. The filter is started with incorrect initial values but converges quickly

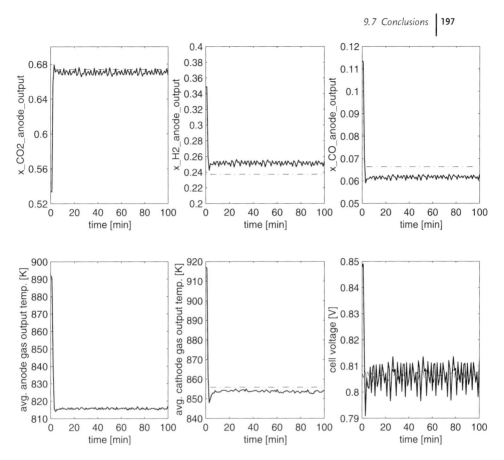

**Fig. 9.4** Convergence of the extended Kalman filter towards a measured state of the HotModule; in the first row: measured (dash-dotted lines) and estimated (solid lines) molar fraction of carbon dioxide, hydrogen and carbon monoxide at the output of the anode gas channel; in the second row: measured (dash-dotted lines) and estimated (solid lines) measurement variables; from left to right: average anode gas output temperature, the average cathode gas output temperature and the cell voltage.

and with good precision towards the measurement and the concentration values. The two load changes are tracked with acceptable accuracy and only small stationary errors remain.

## 9.7
## Conclusions

High temperature fuel cell stacks may show a complex process behaviour due to the high degree of process integration. While currently most high temperature fuel cell stacks are operated manually or via simple controller schemes based on rather crude models, a more advanced process control may help to enhance this

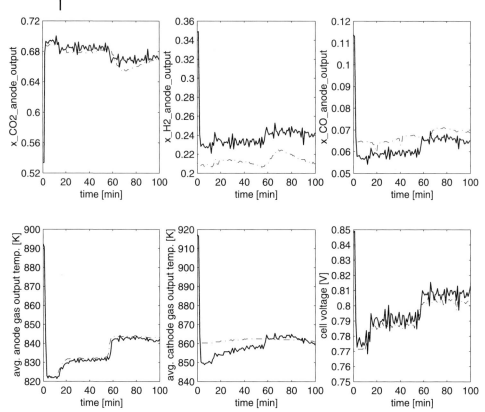

**Fig. 9.5** Convergence of the extended Kalman filter if a measured load change of the HotModule is used as a reference; in the first row: measured (dash-dotted lines) and estimated (solid lines) molar fraction of carbon dioxide, hydrogen and carbon monoxide at the output of the anode gas channel; in the second row: measured (dash-dotted lines) and estimated (solid lines) measurement variables; from left to right: average anode gas output temperature, the average cathode gas output temperature and the cell voltage.

technology, e.g. to obtain higher current and power densities under safe operation conditions. As an example for model-based process control, a state estimator for the HotModule is presented here. The state estimator uses a nonlinear reduced process model of low order derived from a spatially two-dimensional physical reference model. Despite the limited measurement information available, the filter shows good convergence in simulation as well as in experimental tests. The filter gives accurate estimates of the temperature and concentration profiles inside the cell stack, which are very hard to obtain by direct measurements. Thus, the state estimator allows for monitoring the internal stack temperatures, whose control is a key issue for the process operation.

The state estimator can be seen as a soft sensor that supports the operating personal by additional information on the state of the HotModule. But, of course, the state estimator can also be used as a part of an automatic control system for the cell stack. This is shown in [16], where an inferential control scheme is proposed for the HotModule. The control scheme consists of a feedforward part and a feedback part. The feedforward controller sets the feed flow rate of the fuel, the air number, and the steam-to-carbon ratio to the values determined by Heidebrecht [5] in steady state optimisations. A PID controller adjusts the air number in such a way that the maximum temperature is below a given limit also under transient conditions. A second PID controller sets the temperature difference in the stack by modifying the steam-to-carbon ratio. By this rather simple control scheme it is possible to apply rapid changes of the cell current to the HotModule while avoiding over-temperatures and to strong temperature differences in the stack.

The extended Kalman filter presented here only gives good results if the process model is sufficiently accurate. However, it is also possible to estimate some unknown model parameters in addition to the states of the model. This is demonstrated in [11], where a parameter in the reaction kinetics is estimated using a Luenberger observer. The estimated parameter indicates the activity of the catalyst, i.e. its value is a measure for the degradation of the cell stack. In this sense, a state estimator can also be used for process diagnostics.

## Bibliography

1 J. Baker and P. Christofides. Finite-dimensional approximation and control of non-linear parabolic PDE systems. *International Journal of Control*, 73:439–456, 2000.

2 Joachim Birk. *Rechnergestützte Analyse und Lösung nichtlinearer Beobachtungsaufgaben*. Fortschrittberichte VDI, Reihe 8, Nr. 294. VDI-Verlag Düsseldorf, 1992.

3 B.A. Finlayson. *The Method of Weighted Residuals and Variational Principles*. Academic Press, New York, 1972.

4 Arthur Gelb, editor. *Applied Optimal Estimation*. MIT Press, Cambridge, 1974.

5 P. Heidebrecht. *Analysis and Optimisation of a Molten Carbonate Fuel Cell with Direct Internal Reforming (DIR-MCFC)*. PhD thesis, Otto-von-Guericke-Universität Magdeburg, 2005. Fortschritt-Berichte VDI, Reihe 3, Nr. 826, VDI Verlag Düsseldorf.

6 K. Hoo and D. Zheng. Low-order control-relevant models for a class of distributed parameter systems. *Chemical Engineering Science*, 56:6683–6710, 2001.

7 R. Köhler, K.D. Mohl, H. Schramm, M. Zeitz, A. Kienle, M. Mangold, E. Stein, and E.D. Gilles. Method of lines within the simulation environment DIVA for chemical processes. In A. Vande Wouver, Ph. Saucez, and W.E. Schiesser, editors, *Adaptive Method of Lines*, pages 371–406. Chapman & Hall, London, 2001.

8 M. Loève. *Probability Theory*. Van Nostrand Reinhold, Princeton, 1955.

9 D. Luenberger. *Introduction to Dynamic Systems*. Wiley, New York, 1979.

10 M. Mangold, M. Grötsch, M. Sheng, and A. Kienle. State estimation of a molten carbonate fuel cell by an extended Kalman filter. In T. Meurer, K. Graichen, and E.D. Gilles, editors, *Control and Observer Design for Nonlinear Finite and Infinite Dimensional Systems*. Springer, Berlin, 2005.

11  M. Mangold and M. Sheng. Nonlinear model reduction for a two-dimensional MCFC model with internal reforming. *Fuel Cells*, 4:68–77, 2004.

12  T. Obertopp. *Modellierung, Identifikation und Überwachung sicherheitstechnisch schwieriger Prozesse in mehrphasigen Reaktoren.* PhD thesis, Universität Stuttgart, 2001. Fortschritt-Berichte VDI, Reihe 3, Nr. 701, VDI Verlag Düsseldorf.

13  H.M. Park and D.H. Cho. The use of the Karhunen–Loève decomposition for the modeling of distributed parameter systems. *Chemical Engineering Science*, 51:81–98, 1996.

14  C. Prud'Homme, D.V. Rovas, K. Veroy, and A.T. Patera. Mathematical and computational framework for reliable real-time solution of parameterized partial differential equations. *Mathematical Modelling and Numerical Analysis*, 36(5):747–771, 2002.

15  Ralph Rothfuß and Michael Zeitz. Einführung in die Analyse nichtlinearer Systeme. In Sebastian Engell, editor, *Entwurf nichtlinearer Regelungen*, pages 3–22. Oldenbourg, 1995.

16  M. Sheng, M. Mangold, and A. Kienle. A Strategy for the spatial temperature control of a molten carbonate fuel cell system. *Journal of Power Sources*, 162:1213–1219, 2006.

17  F. Tränkle, M. Zeitz, M. Ginkel, and E.D. Gilles. ProMot: a modeling tool for chemical processes. *Mathematical and Computer Modelling of Dynamical Systems*, 6:283–307, 2000.

# 10
# Optimal Control Strategies

*Kati Sternberg, Kurt Chudej, and Hans Josef Pesch*

Many problems arising in science and technology lead to dynamical processes varying in time and space. If sufficient process knowledge is available, those processes may be modelled by partial differential equations (PDEs). In case that one wishes not only to simulate these processes, i.e., to solve the PDEs (in practice almost always numerically), but also to optimise them, one is confronted with PDE constrained optimisation problems being one of today's greatest challenges in applied mathematics. Hereby, certain degrees of freedom within the mathematical models, often constant or at least merely time dependent, are to be determined in an optimal way, for example by minimising or maximizing a given performance criterion.

In view of the dynamical processes that go off in fuel cells chemical reactions, temperature distributions and gas flow regimes may be controlled. For example the gas flows can be affected by changes of density, speed or pressure of the gas flow conditions at the anode inlet. A temperature distribution may be controlled by the chemical reactions or vice versa, e.g., by external heating or cooling devices. Among the multiplicity of possible controls one looks for those being suitable to optimise the system with respect to a desired performance criterion, such as efficiency or lifespan. Since it is generally impossible to compute an analytical solution for such process models because of their enormous complexity, one must solve the optimisation problem numerically, which is, due to the model complexity, a great challenge at the border of today's mathematical and algorithmic knowledge and computer performance.

Within this chapter we firstly present the simulation scenario, which provides a basis for the optimal control problem to be formulated. With minor changes we refer to the spatially 2D MCFC model in Chapter 3 and focus on a change of the total cell current during a fixed given time interval. Due to the different time scalings of the variables we introduce an appropriate interval splitting. This also allows us to observe changes of variables more detailed. Secondly, we provide the mathematical background for the numerical approach used here. This approach is known as *first discretise, then optimise*. Hereby the partial differential equation

*Molten Carbonate Fuel Cells.* Edited by Kai Sundmacher, Achim Kienle, Hans Josef Pesch,
Joachim F. Berndt, and Gerhard Huppmann
Copyright © 2007 WILEY-VCH Verlag GmbH & Co. KGaA, Weinheim
ISBN: 978-3-527-31474-4

system is transferred to a large scale ordinary differential algebraic equations system (ODAEs) by the well-known method of lines. ODAE constrained optimisation problems can be treated using standard techniques, such as the in-house software package NUDOCCCS. As we will see the dimension of the ODAE system to be solved or more precisely the nonlinear programming problem (NLP) into which this system is transformed presents still limitations for today's NLP solvers because of the extremely high dimension of the resulting finite-dimensional constrained nonlinear optimisation problem. Thirdly, we present numerical results for the optimally controlled load changes. Here, the aim is to drive the cell voltage close to the value of the steady state. We will compare the estimated optimally controlled cell voltage with pure simulation results and show the superiority of the optimal control approach. In particular we look at five different control variables, quantify their influence on the cell voltage and assess their practical suitability for the control purpose under investigation.

## 10.1
## Model and Simulation Setting

The simulation scenario is based on the 2D dimensionless reference model of Chapter 3 with a direct feed into the anode channels. The molar fractions $\chi_{i,a}$ and $\chi_{i,c}$ for $i = \{CH_4, H_2, H_2O, CO, CO_2, O_2, N_2\}$, the gas temperatures $\vartheta_a$ and $\vartheta_c$ as well as the molar flows $\gamma_a$ and $\gamma_c$ in the anode and cathode channel are described by semi-linear and quasi-linear hyperbolic-type transport equations (3.20)–(3.22) and (3.46)–(3.48) with highly nonlinear source terms. Independent variables are the time $\tau$ and the two space variables $(\zeta_1, \zeta_2)$. Moreover, due to the absence of the derivative in time, the PDEs (3.22) and (3.48) are degenerated making the system a system of partial differential algebraic equations (PDAEs). The parabolic PDE (3.65) for the solid temperature $\vartheta_s$ has nonlinear source terms, too. Note that in order to reduce the amount of computation the gas concentrations in the channels and the pores are assumed to be equal. This means $\varphi_{i,ac} = \chi_{i,a}$ and $\varphi_{i,cc} = \chi_{i,c}$ for $i = \{CH_4, H_2, H_2O, CO, CO_2, O_2, N_2\}$. Therefore boundary conditions at the anode inlet are necessary. For control purposes they can be chosen as time dependent control functions, the so-called boundary controls. In contrast, the boundary conditions at the cathode inlet are given via the solution of a system of algebraic equations (3.32)–(3.34) associated with the catalytic combustion and ordinary differential equations (3.38)–(3.40) associated with the reversal chamber. These equations include integral terms averaging the anode outlet. Additionally, the electrical potentials $\Phi_a^L$, $\Phi_c^L$, and $\Phi_c^S$ have to be included in the model, see Eqs. (3.73)–(3.75). These differential equations include integral terms in the right hand sides and exhibit no space derivatives even though $\Phi_a^L$ and $\Phi_c^L$ are space dependent. The standard initial and parameter values can be found in the appendix. For the sake of simplicity we have not included a (partial) feedback of the cathode exhaust to the catalytic burner for the purpose of optimisation and thus $R_{back} = 0$ is to be chosen.

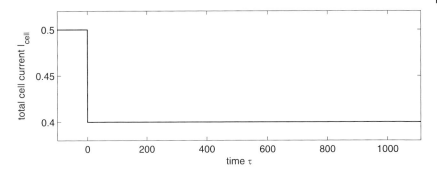

**Fig. 10.1** Simulated load change.

For the subsequent optimal control of the fuel cell the following dynamic simulation scenario is chosen [6]: Over the time interval $\tau \in [0, 1111.1]$ a load change is simulated. We have placed the jump of the cell current at time $\tau = 0$ with $I_{cell} = 0.5$ for $\tau \leq 0$ and $I_{cell} = 0.4$ for $\tau > 0$; see Fig. 10.1.

Due to the different time scales of the components the entire time interval is divided into five subintervals. Note that the cell voltage can vary by a considerable amount within milliseconds, whereas the material flow quantities need seconds or minutes, respectively. The solid temperature lives on a largest time scale. It is given by minutes to hours. Within these ranges the individual components may achieve steady state. Therefore we focus on the time intervals $[0, 0.1]$, $[0.1, 1.1]$, $[1.1, 11.1]$, $[11.1, 111.1]$, and $[111.1, 1111.1]$. Figure 10.2 shows the cell voltage during load change. The jump of the cell voltage in the first time interval $[0, 0.1]$ can be clearly seen. This means the voltage is affected within milliseconds after the load change. Subsequently, the cell voltage increases over the intervals $[0.1, 1.1]$, $[1.1, 11.1]$, and $[11.1, 111.1]$ due to the change of the gas flows. It ap-

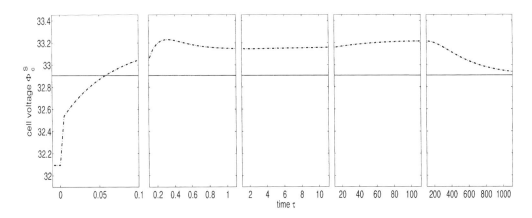

**Fig. 10.2** Behaviour of cell voltage during load change.

proaches steady state in the last interval only, when the solid temperature approaches its stationary value, too.

More simulation results of a load change can be found in [1–3].

## 10.2
## Mathematical Methods

The PDAE system under consideration can be written in compact form by

$$A\frac{\partial x}{\partial \tau} = B\Delta x + C_1 \frac{\partial x}{\partial \zeta_1} + C_2 \frac{\partial x}{\partial \zeta_2} + f(x) \tag{10.1}$$

Hereby $A$ is a singular diagonal matrix containing only zeros and ones. The matrix $B = \text{diag}(\lambda, 0, \ldots, 0)$ is diagonal, too, if we choose the vector $x$ containing all the unknown functions of the equations in Section 10.1 to be $x := (\vartheta_s, \ldots, \Phi_c^S)$. Here the scaled Laplacian has to be understood componentwise. The singular diagonal matrices $C_1$, respectively $C_2$, depend on the auxiliary variables $v_a := \gamma_a \vartheta_a$, respectively $v_c := \gamma_c \vartheta_c$, making the system quasi-linear. Finally the term $f(x)$ contains all other nonlinear dependences.

In order to solve this complex nonstandard PDAE constrained optimisation problem numerically we have chosen the approach *first discretise, then optimise* in contrast to *first optimise, then discretise*. The latter approach means that optimality conditions are firstly derived in a function spaces context which are discretised afterwards. We however favour the first approach. Due to the various model updates during the development of the hierarchy of MCFC models (which is typical when dealing with real world problems) the latter approach is not applicable, although it is considered to be mathematically more safeguarded.

Thus, we apply the (vertical) method of lines as the method of choice. Herewith we obtain a large scale ODAE system of the form:

$$A_h \frac{\partial x_h}{\partial \tau} = g(x_h) \tag{10.2}$$

for which standard techniques can be used, if the dimension of the nonlinear programming problems, to which the ODAE constrained optimisation problem is transcribed, remains moderately large.

The numerical approximation $x_h(\tau)$ of the exact solution $x(\tau, \zeta_1, \zeta_2)$ is accordingly obtained by semi-discretisation in the spatial variables $\zeta_1$ and $\zeta_2$. Hereby, the scaled Laplace operator is approximated by the five point difference star. For the convection terms upwind formulas are used taking into account the known wind direction in the anode and cathode gas channels, respectively. Despite the complicated boundary conditions they can be handled in the usual way like Dirichlet and Neumann conditions the values of which are partly obtained via the solution of an ODAE system.

In Eq. (10.2), $A_h$ is a constant singular diagonal matrix containing again only zeros and ones. The nonlinear function g stands for the discretised right-hand side of (10.1).

Now we are going to present the numerical simulation results. The software package NUDOCCCS due to Büskens [4] is designed for the numerical solution of relatively large scale ODE and DAE constrained optimal control problems. The numerical integration of the DAE system within NUDOCCCS is performed by the fifth order implicit Runga–Kutta method RADAU 5.

## 10.3
## Optimal Control of a Load Change

One aim of the optimal control in the case of load changes is to attain the steady state value $\Phi_{\text{refer}}$ of the cell voltage as fast as possible after the load change. The reference cell voltage $\Phi_{\text{refer}} = 32.9041986$ was here determined by the simulation with the default values of the appendix and $I_{\text{cell}} = 0.4$. Therefore the deviation of the cell voltage to the desired value, i.e., $(\Phi_{\text{c}}^{\text{S}}(\tau) - \Phi_{\text{refer}})^2$, summed up over the entire time interval, yields the objective function to be minimised. Problems of this type are known as tracking problems. As controls the air supply $u_{\lambda_{\text{air}}}$ at the burner, the amount of methane $u_{\chi_{\text{a, in, CH4}}}$ at the anode inlet, the flow density $u_{\Gamma_{\text{a, in}}}$ at the anode inlet, the temperature $u_{\vartheta_{\text{air}}}$ of the supplied air at the burner and the temperature $u_{\vartheta_{\text{a, in}}}$ of the supplied gas at the anode inlet are chosen. The controls must obey the following bounds defining the set $U_{\text{ad}}$ of admissible controls: $u_{\lambda_{\text{air}}} \in [2.1, 2.3]$, $u_{\chi_{\text{a, in, CH4}}} \in [0.29, 0.31]$ with $\chi_{\text{a, in, H}_2\text{O}} = 1 - u_{\chi_{\text{a, in, CH4}}}$, as well as $u_{\Gamma_{\text{a, in}}} \in [0.9, 1.1]$, $u_{\vartheta_{\text{air}}} \in [1.4, 1.6]$, and $u_{\vartheta_{\text{a, in}}} \in [2.9, 3.1]$. All control variables are merely time dependent. We have a so-called boundary control problem in contrast to distributed controls which enter the right hand side of the differential equations and being space dependent. Additionally, regularisation terms with the weighting factor $\mu = 0.5$ taking into account the control costs are included into the objective function. Such regularisation terms smooth the control behaviour. Thus the objective functions is modelled by

$$\min_{u \in U_{\text{ad}}} \int_0^{1111.1} \left[ (\Phi_{\text{c}}^{\text{S}}(\tau) - \Phi_{\text{refer}})^2 + \mu(u_{\lambda_{\text{air}}} - \lambda_{\text{air}})^2 + \mu(u_{\chi_{\text{a, in, CH4}}} - \chi_{\text{a, in, CH4}})^2 \right.$$

$$\left. + \mu(u_{\Gamma_{\text{a, in}}} - \Gamma_{\text{a, in}})^2 + \mu(u_{\vartheta_{\text{air}}} - \vartheta_{\text{air}})^2 + \mu(u_{\vartheta_{\text{a, in}}} - \vartheta_{\text{a, in}})^2 \right] d\tau$$

Again the time interval $[0, 1111.1]$ is divided into five subintervals, whereby the state of the fuel cell at the exit point of a time interval yields the initial state for the next time interval. In each time interval the number of discrete time steps is set to 21. Hence, the controls are optimised over the subintervals only. Afterwards the optimal controls on each subinterval are joined together into a suboptimal control over the whole time interval $[0, 1111.1]$. An optimal control over the whole

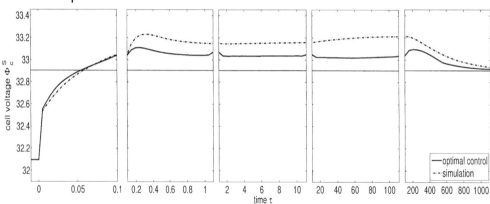

**Fig. 10.3** Optimal controlled and simulated cell voltage during a load change.

interval cannot be computed due to the enormous dimension of the nonlinear programming problem that would arise if doing so.

An advantage of our approach is that the dynamic behaviour in the first intervals can be resolved in more details. Figure 10.3 shows the optimally controlled cell voltage over the five time intervals in comparison to the simulation. The voltage keeps its characteristic behaviour with the jump directly after the load change at $\tau = 0$, which is then followed by an increase beyond the value of the steady state and a decrease to the stationary value. However, in comparison to the simulation the jumps are attenuated and the value of the steady state is reached much earlier. The optimally controlled cell voltage differs from time $\tau \approx 800$ in less than 0.1% from the reference value $\Phi_{\text{refer}}$, whereas the simulated cell voltage falls below this level only at $\tau \approx 1111$.

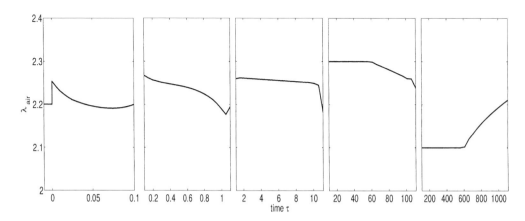

**Fig. 10.4** Optimal air inlet $u_{\lambda_{\text{air}}}$ at the catalytic combustion.

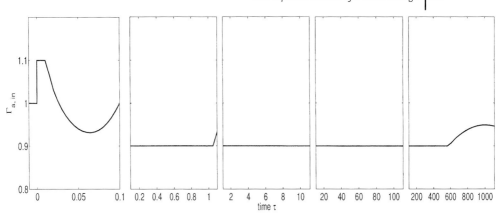

**Fig. 10.5** Optimal molar flow $u_{\Gamma_{a,in}}$ at the anode inlet.

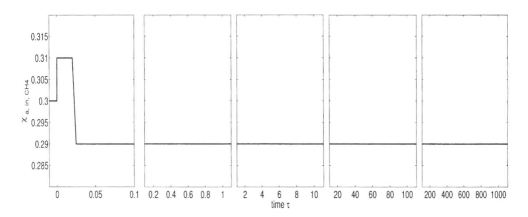

**Fig. 10.6** Optimal methane inlet $u_{\chi_{a,in,CH4}}$ at the anode inlet.

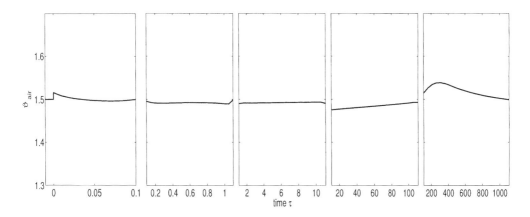

**Fig. 10.7** Optimal air temperature $u_{\vartheta_{air}}$ at the catalytic combustion.

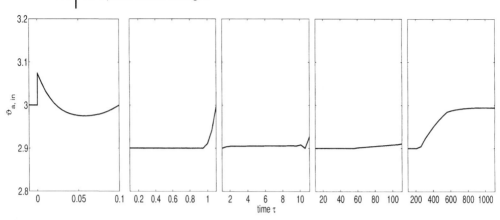

**Fig. 10.8** Optimal gas temperature $u_{\vartheta_{a,in}}$ at the anode inlet.

Figures 10.4 to 10.8 show the optimal controls. Except at the interval boundaries the controls are smooth. Of particular interest are the controls $u_{\lambda_{air}}$ for the
air inlet at the catalytic combustion and $u_{\Gamma_{a,in}}$ for the molar flow at the anode inlet. These two controls significantly influence the cell voltage. For this purpose we
compare the cell voltage controlled by only these two controls with the one controlled by the five controls and with the simple simulation; see Fig. 10.9. In particular the molar flow is responsible for the short-time behaviour in the first four
time intervals, whereas the air inlet accelerates the faster adjustment to the steady
state value in the last time interval.

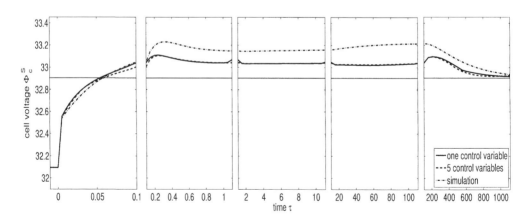

**Fig. 10.9** Cell voltage with two control variables $u_{\lambda_{air}}$ and $u_{\Gamma_{a,in}}$,
compared to the one with the five control variables and the one based
on pure simulation.

## 10.4
## Summary and Conclusion

In this chapter we have presented a so-called load change, a change of the total cell current during a fixed given time interval, which provides the simulation scenario for the optimal control problem. With minor changes we referred to the model in Chapter 3 describing the processes in the MCFC using a large scale partial differential algebraic equation system with integral terms in the right hand side together with its nonstandard boundary conditions including an ODE system. Due to the chosen concept *first discretise, then optimise* the applied semi-discretisation yields a very large scale DAE system, which was solved using standard techniques like the software Nudocccs. Concerning the different time scaling of the variables we have introduced interval splitting, which allows a more detailed observation of the variables. The aim of the optimal control was to attain the value of the cell voltage in steady state as soon as possible after the load change.

The numerical results of the optimal control show a good performance and a fast achievement of this objective. Compared to the simulation the deviation of the cell voltage to the desired steady state value is smaller in each time step of the observed time interval. Furthermore the cell voltage reaches the steady state value much earlier. In fact, in the simulation one achieves a cell voltage, that deviates 0.1% from the steady state value at the end of the observed time interval. In optimisation we only need 3/4 of the time interval to achieve this 0.1% deviation. Among the considered boundary controls we constituted the optimal control of the molar flow at the anode inlet and the air supply to the catalytic combustion, as those controls that are mainly responsible for the long and the short-time behaviour of the cell voltage.

The obtained numerical results are of practical interest, since they allow to accelerate load changes during the operation of the fuel cell. At the moment a load change is very extensive and therefore expensive, because the cell voltage is step-wisely adapted in a tedious procedure. A special case of this application is the start up of the fuel cell, which takes hours and days.

Moreover, this approach enables further investigations with respect to parametric sensitivity analysis, which is concerned with the behaviour of the optimal solution due to parameter perturbations, e.g. some changes in the gas flow ([5]). Sensitivity analysis indicates how to handle possibly inexact parameter values in real time. For those purposes we can use the same mathematical model as well as the same software Nudocccs.

### Bibliography

1 Heidebrecht, P., Modelling, Analysis and Optimisation of a Molten Carbonate Fuel Cell with Direct Internal Reforming (DIR-MCFC). VDI Fortschritt Berichte, Reihe 3, Nr. 826, VDI Verlag, Düsseldorf, 2005.
2 Pesch, H.J., Sternberg, K., Chudej, K., Towards the Numerical Solution of a

Large Scale PDAE Constrained Optimization Problem Arising in Molten Carbonate Fuel Cell Modeling. In: G. Di Pillo, M. Roma (Eds.), Large Scale Nonlinear Optimization, Springer, Berlin, 2006, 243–253.

3 Chudej, K., Sternberg, K., Pesch, H.J., Simulation and Optimal Control of Molten Carbonate Fuel Cells. In: I. Troch, F. Breitenecker (Eds.) Proc. of 5th MATHMOD Vienna, ARGESIM Report no. 30, ARGESIM-Verlag, Wien, 2006.

4 Büskens, C., Optimierungsmethoden und Sensitivitätsanalyse für optimale Steuerprozesse mit Steuer- und Zustands-Beschränkungen. Dissertation, Universität Münster, 1998.

5 Sternberg, K., Simulation, Optimale Steuerung und Sensitivitätsanalyse einer Schmelzkarbonat-Brennstoffzelle mithilfe eines partiellen differential-algebraischen dynamischen Gleichungssystems. Dissertation, Universität Bayreuth, 2007.

6 Sternberg, K., Chudej, K., Pesch, H.J., Suboptimal Control of a 2D Molten Carbonate Fuel Cell PDAE Model. *Mathematical and Computer Modelling of Dynamical Systems*, to appear.

# 11
# Optimisation of Reforming Catalyst Distribution

*Peter Heidebrecht and Kai Sundmacher*

## 11.1
### Introduction

Although MCFC systems exist and are operational today in many different fields of applications, their design and operating conditions are determined by empirical knowledge and experience rather than by strict optimisation. In this chapter, the usefulness of a model-based optimisation is demonstrated. The optimisation does not consider the full model from Chapter 3, but an idealised system. According to the conceptual studies in Chapter 8 a pure DIR–MCFC would be an ideal MCFC system in a sense that a maximum conversion can be obtained with minimal technical equipment. The following optimisations apply an MCFC model without the IIR channels, where a mixture of methane and water is directly fed into the anode channels. For the sake of clarity, although the optimisations are conducted with a dimensionless model, dimensional values will be given wherever possible.

The optimisations have to consider several aspects with respect to a safe and efficient operation of the fuel cell. As already mentioned in previous chapters (Chapters 3 and 6), temperature is the most vital state in a high temperature fuel cell like the MCFC. In order to allow the use of inexpensive materials, the temperature may not be too high at any location in the cell. Also, high temperature leads to fast degradation of the nickel catalyst. On the other hand, too low temperatures lead to high polarisation losses due to the Arrhenius effect, which slows down the reaction rates and the ion transport through the electrolyte. In a DIR cell, a mixture of methane and water enters the anode channel, causing a very high reforming reaction rate near the inlet region. This leads to a significant heat sink and thereby a possible cold spot in that region. This cold spot leads to low reaction rates of the electrochemical process. Also, the hydrogen production is insufficient at low temperatures. On the other hand, close to the anode outlet region, the reforming process has almost come to full conversion, which means that the only way to bleed off heat is by convective transport in the gas phases. In this region, the highest temperature is likely to occur.

*Molten Carbonate Fuel Cells.* Edited by Kai Sundmacher, Achim Kienle, Hans Josef Pesch, Joachim F. Berndt, and Gerhard Huppmann
Copyright © 2007 WILEY-VCH Verlag GmbH & Co. KGaA, Weinheim
ISBN: 978-3-527-31474-4

Another important effect in high temperature fuel cells is the deactivation of the catalyst by carbon deposition. This reversible effect can quickly deactivate the catalyst in the anode channel and thereby render the cell useless. Although this is a more significant problem for solid oxide fuel cells (SOFC) due to temperature level, it can also occur in MCFCs. To avoid carbon deposition, sufficient amounts of water have to be supplied with the feed gas.

Besides the desired oxidation of hydrogen and carbon monoxide and the reduction of oxygen, other electrochemical reactions may occur in an MCFC. If the cell voltage is too low, the nickel catalyst dissolves in the electrolyte in ionic form, which also leads to a quick deactivation of the catalyst.

## 11.2
## Objective Functions and Optimisation Parameters

A fuel cell can be optimised in many ways with different objective functions. For example, the minimisation of the maximum temperature difference in a stack requires a simple objective function, but economical objective functions which aim at the maximisation of monetary benefit tend to become rather complex for optimisation purposes. In our case, a purely technical objective function is chosen, which is the electric efficiency of the system. It is defined as the ratio of the electric cell power, $P_{cell}$, diminished by the system's parasitic power consumption divided by the lower heating value of methane in the anode feed. The system's power consumption is mainly determined by the power demand of the large air blowers that circulate the cathode gas flow, $P_{blower}$. For simplification, a proportional correlation between cathode gas flow and the blower power is assumed. The enthalpy of the feed gas is the product of the total molar feed flow, $G_{a,in}$ and the molar fraction of methane in the feed, quantified by the steam-to-carbon ratio, $S/C$, and the molar standard enthalpy of combustion of methane, $\Delta_c h^{\theta}_{CH_4}$. Thus, the objective function reads as follows:

$$\eta_{el} = \frac{P_{sys}}{H_{a,in}} = \frac{P_{cell} - P_{blower}}{G_{a,in}} \cdot \frac{1}{1 + S/C \cdot (-\Delta_c h^{\theta}_{CH_4})} \tag{11.1}$$

Two different types of optimisations are performed with this objective function: First, only the input parameters are used to optimise the objective function for a given average current density, $i_{cell}$:

$$\eta_{el}(G_{a,in}, S/C, \lambda_{air}, R_{back}) \rightarrow \max! \tag{11.2}$$

The optimisation parameters and the output information from the model are depicted in Fig. 11.1. This optimisation yields optimal operating conditions and repeating this optimisation for several different values of average current density results in several sets of optimal operating parameters and a current–voltage curve under optimal conditions.

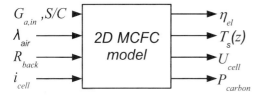

**Fig. 11.1** Relevant input and output variables of the 2D cross flow MCFC model.

In a DIR–MCFC, the reforming reaction is a spatially distributed cooling process. Coordinating this cooling effect with the heat sources from the electrochemical reactions could help to improve the performance of the cell. Manipulation of the reaction rates can be achieved by spatially distributed catalyst activities or catalyst densities. While this is rather complicated to realise for the electrochemical reactions, it is fairly easy to do with the reforming catalyst, which is applied into the channels independently from the electrodes. Consequently, the second optimisation task includes the spatial distribution of the reforming catalyst inside the anode channel. For the purpose of this optimisation, the overall 2D cell area is divided into four sections each with a constant catalyst density. Starting from the so-called 'base case' configuration with a completely constant catalyst density distribution, one factor, $F_{Da}$, in each section is used to describe the change in catalyst density in that section. As the amount of catalyst that can be placed in the anode channel is limited, this factor is also limited to a maximum value of 3.0. Thus the second optimisation uses four additional optimisation variables:

$$\eta_{el}(G_{a,in}, S/C, \lambda_{air}, R_{back}, F_{Da,1}, F_{Da,2}, F_{Da,3}, F_{Da,4}) \rightarrow \max! \tag{11.3}$$

The number of four sections was chosen for two reasons: firstly in order to keep the numerical effort in an acceptable range. Using the results from this optimisation as an initial guess, one could probably increase the number of sections to $9\,(3 \times 3)$ sections. As the numerical effort increases rapidly with the number of optimisation parameters, the number of sections should not be much higher. The second reason is that very fine distributions are hardly applicable from a practical point of view.

Both optimisation tasks consider several constraints. Equations (11.4) and (11.5) express the fact that the cell temperature is limited to a certain window. This is to avoid material damage due to too high temperature and to guarantee a minimum temperature necessary to ascertain high reaction rates and a good ion conductivity in the electrolyte. The third constraint (Eq. (11.6)) limits the maximum temperature difference in order to avoid mechanical stress and leakages due to different thermal expansions. The fourth constraint (Eq. (11.7)) accounts for a minimum cell voltage, below which undesired side reactions of the nickel catalyst with the carbonate melt occur. The last constraint (Eq. (11.8)) takes into account the possibility of carbonisation. The criterion ensures that carbonisation is thermodynamically impossible at any location in the cell:

$$\min(T_s(z_1, z_2)) = T_{s,\min} \geq T_{\lim,\min} = 591 \ ^\circ C \tag{11.4}$$

$$\max(T_s(z_1, z_2)) = T_{s,\max} \leq T_{\lim,\max} = 681 \ ^\circ C \tag{11.5}$$

$$\max(T_s(z_1, z_2)) - \min(T_s(z_1, z_2)) = \Delta T_{s,\max} \leq \Delta T_{\lim,\max} = 60 \ ^\circ C \tag{11.6}$$

$$U_{cell} \geq U_{\lim,\min} = 694 \ mV \tag{11.7}$$

$$\exists j : \Delta_{Rg}C_j(T_a(z_1, z_2), x_{i,a}(z_1, z_2)) \geq 0 \tag{11.8}$$

In addition to this, the cathode reflux ratio, $R_{back}$, is limited to a maximum value of 0.5. The full set of discretised steady state equations of the 2D MCFC model are considered as equality constraints to this optimisation.

## 11.3
## Numerical Aspects

The spatially distributed MCFC model from Chapter 3 is used for these optimisations. It is implemented in ProMoT (Process Modelling Tool, [1]) and solved within the simulation environment DIVA [2]. The sequential quadratic programming (SQP) algorithm E04UCF from the NAG library [3], offered by the DIVA package, is employed. The model is in transient formulation, because directly solving the steady state equation system requires excellent starting values and is thus not preferable. Therefore, a dynamic approach is chosen, in which the equation system is integrated for every optimisation step until steady state is reached. Then, the objective function is evaluated at steady state. The highly effective SQP optimisation algorithm reaches the optimum within only a few steps, which is due to the use of parameter sensitivities. These are delivered by the underlying integration algorithm. To work properly, the integrator requires a continuously derivable equation system including the variables for the constraints. This is not the case with the minimum and maximum temperature in Eqs. (11.4) to (11.6) and with the carbonisation criterion in Eq. (11.8). As long as, say, the lowest temperature stays at the same position in the cell, $T_{s,\min}$ is continuously derivable. But as soon as the location of the lowest temperature changes, the derivability is no longer given and the integrator fails.

Thus the following two differential equations are introduced to provide continuously derivable extremal temperatures:

$$\frac{dT_{s,\min}}{dt} = \frac{1}{\tau_\vartheta} \cdot (\min(T_s(z_1, z_2) - T_{s,\min})) \tag{11.9}$$

$$\frac{dT_{s,\max}}{dt} = \frac{1}{\tau_\vartheta} \cdot (\max(T_s(z_1, z_2) - T_{s,\max})) \tag{11.10}$$

where $\tau_\vartheta$ is an arbitrary time constant which should be significantly lower than the slowest time constant in the fuel cell model. With this, also the maximum temperature difference in Eq. (11.6) is derivable.

Also, the carbon deposition criterion requires attention. Here, a kind of penalty function is applied. Consider the Gibbs energy of one of the carbonisation reactions $j$ at a certain location $z$ and a certain time $t$. Then, a continuous, limited, reaction related penalty function is defined according to

$$
p_j(z,t) = \begin{cases} 0, & \text{if } \Delta_r g_{Cj}(z,t) \geq 0 \\ \frac{(\Delta_R g_{Cj}(z,t))^2}{1+(\Delta_R g_{Cj}(z,t))^2}, & \text{if } \Delta_r g_{Cj}(z,t) < 0 \end{cases} \tag{11.11}
$$

This function is zero if the specific carbonisation reaction $j$ is not likely to happen and it continuously increases to one if the reaction is likely to happen. Carbonisation will definitely happen if all considered carbonisation reactions have a negative Gibbs enthalpy at the respective position. This means that each of them will run in forward direction, i.e. it will deposit carbon, and not a single reaction will run in backward direction, i.e. consume deposited carbon. With this in mind, the local penalty function of carbonisation is defined as:

$$
p(z,t) = \prod_j p_j(z,t) \tag{11.12}
$$

This function is zero if at least one carbonisation reaction runs in backward direction. Otherwise it will assume a positive value between zero and one.

Because carbonisation is not allowed to occur at any location in the cell, the overall penalty function for carbonisation (or overall carbonisation criterion) is expressed as the integral of the local penalties:

$$
P(t) = \int_z p(z,t)\,dz \tag{11.13}
$$

The reaction specific penalty functions depend on continuously derivable states like concentrations and temperature. The overall carbonisation criterion is comprised of these by continuous operations, thus this formulation is applicable for the integration algorithm. The constraint (Eq. (11.8)) now reads:

$$
P(t) \leq 0 \tag{11.14}
$$

With these reformulations of the inequality constraints it is possible for the integrator to provide the required sensitivities to the optimisation algorithm. A typical optimisation run with an $8 \times 8$ spatial discretisation grid includes about 1200 ODEs, which are repeatedly integrated until steady state, five inequality constraints and four optimisation parameters. It takes between one and three hours of CPU time on a standard 1 GHz machine.

## 11.4
## Results

### 11.4.1
### Optimisation of Input Conditions at Constant Catalyst Density

The optimisation of the first mentioned type (Eq. (11.2)) is performed at various average cell current densities. The result is a set of operating conditions for each cell current which fulfils all constraints and offers an optimal electric efficiency. These points form an optimal current–voltage curve (Fig. 11.2). The circles in Fig. 11.3 show the corresponding electric efficiencies. They range from more than 57% at low currents to 38% at high cell currents. Note that these numbers differ from those in Chapter 5, as here we consider a pure DIR–MCFC. For more detailed information about the optimisation results, we refer to [4]. Figure 11.4 shows the temperature field inside the cell at base case condition, that is at $i_{cell} = 124.6$ mA/cm$^2$. The highest temperature occurs at the outlet/outlet corner, while the lowest temperature is found at the inlet/inlet corner of anode and cathode channels. This will be of interest later on.

From the details of the optimisation results, operating regimes can be identified. At low current densities, for example, the temperature is low and so the air number, $\lambda_{air}$, is also low because an increase of the air flow would decrease the cell temperature below the critical value; at high current densities, the air number is very high in order to cool the cell down. Similar mechanisms can be identified concerning the amount of gas feed and other parameters.

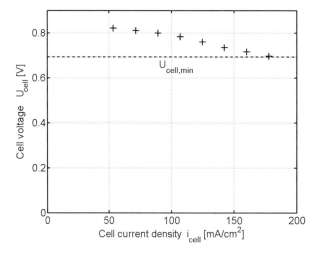

**Fig. 11.2** Current–voltage curve under optimal operating conditions for a system with constant catalyst distribution.

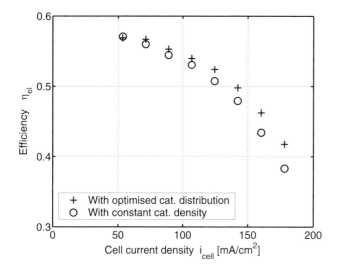

**Fig. 11.3** Effect of optimisation of the reforming catalyst distribution; electric system efficiency over cell current for a system with constant reforming catalyst distribution and for a system with optimised catalyst distribution.

Although the different load cases are dominated by different parameters, one common feature stand out: at each optimal load case, the temperature difference across the cell is at its maximum allowed value. In other words, each optimum is limited by the maximum temperature difference constraint.

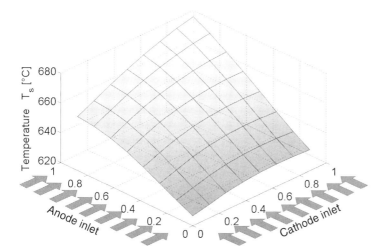

**Fig. 11.4** Cell temperature at $i_{cell} = 124.6$ mA/cm$^2$ with optimised operating conditions and constant reforming catalyst density.

11.4.2

**Optimisation of the Reforming Catalyst Density Distribution**

Because the temperature difference proves to be the limiting constraint in the previous optimisations, the spatial reforming catalyst distribution shall now be used to homogenise the temperature profile, thereby relaxing this constraint and allowing for higher efficiencies. This corresponds to the second type of optimisation (see Eq. (11.3)). Because the catalyst density cannot be changed during operation, this optimisation is only performed once at a base case current density ($i_{cell} = 124.6$ mA/cm$^2$). The result is not only a set of optimal operating conditions, but also four factors describing the optimal reforming catalyst density distribution in the anode channel, as illustrated in Fig. 11.5.

The anode channels on the right-hand side have a slightly increased reforming catalyst density near their inlet (section 1). This leads to a fast hydrogen generation in that section, providing a certain amount of hydrogen for the oxidation reaction. The result is a moderate cooling effect and a high current density. The relatively depleted gas then enters the channels second half (section 3), where a maximum amount of reforming catalyst converts the remaining methane to hydrogen as quickly as possible.

In the anode channels on the left-hand side, the situation is different. Near the inlet, the reforming catalyst density is very low (section 2). This section is not temperature critical (see Fig. 11.4), so the catalyst density here is a compromise between the need for hydrogen for the oxidation reaction and the need to spare the cooling effect of the reforming reaction for the hot area in the second half of these channels. In this configuration, sufficient methane is left in the gas when it enters section 4, where it is quickly reformed. This provides a significant cooling

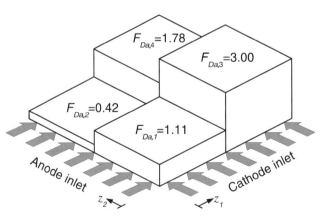

**Fig. 11.5** Result of the optimisation of the reforming catalyst density at base case, i.e. $i_{cell} = 124.6$ mA/cm$^2$; optimal configuration of catalyst density factors.

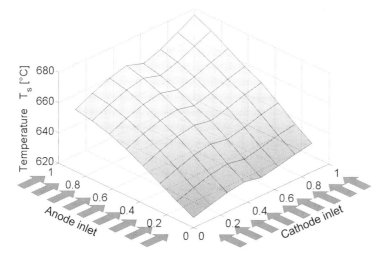

**Fig. 11.6** Cell temperature at $i_{cell} = 124.6$ mA/cm$^2$ with optimised reforming catalyst density distribution and optimised operating conditions.

effect in the hot area of the cell. With a reduced maximum temperature, the other input parameters can be adjusted in such a way that the system efficiency is increased, until the maximum allowable temperature difference is reached again.

Unlike in the right-hand side anode channels, the catalyst density in the channels second half is not at its maximum value (section 4). Increasing the catalyst density in this section would lead to a higher current density here. But then, as the total cell current is constant, current density in the section 1 would decrease. Thereby, an important heat source would be diminished and the cell temperature in that section would decrease and thus the allowable temperature difference would be exceeded. The cell temperature distribution for optimised catalyst density is plotted in Fig. 11.6.

### 11.4.3
### Optimisation of the Input Conditions for a System with Optimised Catalyst Density

Using this catalyst distribution, a new optimal current–voltage curve is calculated. Figure 11.3 shows the results of these optimisations and compares them against those of the system with constant catalyst distribution. It turns out that the catalyst distribution increases the system efficiency not only at the point for which it has been optimised, but for all the medium and high current density load cases. Only at low average cell currents, its performance is equivalent to the system with constant configuration. From an economic point of view, low currents are not attractive, so these operating points are of minor importance.

## 11.5
## Summary

In this chapter, two optimisations for a DIR–MCFC are presented. The optimisation of input parameters for an MCFC is successfully performed based on the spatially two-dimensional cross-flow model. The operating conditions obtained from this optimisation not only yield optimal efficiency at any given cell current, but they also guarantee safe operation of the system with regard to temperature constraints, cell voltage and carbon deposition.

In a system mainly governed by temperature effects, the second optimisation balances the cooling effect of the reforming reaction with the heat source from the electrochemical reactions by optimising the reforming catalyst density distribution. With the catalyst distribution obtained at one single operating point, the system performance is significantly increased over the whole range of cell currents, especially at high currents. This strongly indicates that this measure is advisable in future DIR–MCFC systems.

Extensions of these optimisations are possible. Firstly, the objective function could involve economical aspects instead of a purely technical electrical efficiency. Secondly, the model can be extended to incorporate other reforming stages like the IIR and the ER. It is clear that with an extended model, say with IIR as in Section 3.1, the options for an optimisation of design parameters significantly broaden. For example, not only the spatial distribution of the Damkoehler numbers in the IIR and the DIR are of interest, but also the relative size of these channels could be included in the optimisation. Furthermore, optimisation results should be analysed with respect to the sensitivity towards uncertain parameters like the Stanton numbers or the electrochemical Damkoehler numbers.

### Bibliography

1 Tränkle, F., Zeitz, M., Ginkel, M., Gilles, E.D., PROMOT: A Modeling Tool for Chemical Processes, Mathematical and Computer Modelling of Dynamical Systems 6 (3), 2000, 283–307.

2 Köhler, R., Mohl, K., Schramm, H., Zeitz, M., Kienle, A., Mangold, M., Stein, E., Gilles, E.D., in Vande Wouver, A., Saucez, P. and Schiesser, W. (Eds.), Adaptive Method of Lines,

Chapman & Hall/CRC Press, Boca Raton, 2001.

3 NAG Ltd., NAG Fortran Library Document, Oxford, England, 1993.

4 Heidebrecht, P., Sundmacher, K., Optimisation of reforming catalyst distribution in a cross-flow molten carbonate fuel cell with direct internal reforming (DIR-MCFC), Industrial & Chemical Engineering Chemistry Research 44, 2005, 3522–3528.

# Appendices

*Molten Carbonate Fuel Cells.* Edited by Kai Sundmacher, Achim Kienle, Hans Josef Pesch,
Joachim F. Berndt, and Gerhard Huppmann
Copyright © 2007 WILEY-VCH Verlag GmbH & Co. KGaA, Weinheim
ISBN: 978-3-527-31474-4

# A
# List of Symbols

**Table A.1** Dimensionless variables and parameters

**Latin symbols**

| | | |
|---|---|---|
| $Arr_j$ | $= \dfrac{\text{Activation energy}}{\text{Characteristic sensible energy}}$ | Arrhenius number of reaction $j$ |
| $c$ | $= \dfrac{\text{Charge capacity}}{\text{Characteristic charge capacity}}$ | Surface related charge capacity |
| $c_{p,i}$ | $= \dfrac{\text{Molar heat capacity}}{\text{Characteristic heat capacity}}$ | Molar heat capacity of component $i$ |
| $c_{p,s}$ | $= \dfrac{\text{Solid heat capacity}}{\text{Characteristic heat capacity}}$ | Heat capacity solid phase |
| $\bar{c}_p$ | $= \dfrac{\text{Molar heat capacity}}{\text{Characteristic heat capacity}}$ | Molar heat capacity of gas mixture |
| $Da_j$ | $= \dfrac{\text{Characteristic reaction rate}}{\text{Characteristic convective flow rate}}$ | Damkoehler number of reaction $j$ |
| $D_i$ | $= \dfrac{\text{Characteristic mass exchange density}}{\text{Characteristic convective flow rate}}$ | Mass transport coefficient in pores of component $i$ |
| $F$ | $= \dfrac{\text{Characteristic feed rate}}{\text{Characteristic current}}$ | Dimensionless Faradaic constant |
| $I_{cell}$ | $= \dfrac{\text{Cell current}}{\text{Characteristic cell current}}$ | Total cell current |
| $i$ | $= \dfrac{\text{Current density}}{\text{Characteristic current density}}$ | Surface cell current density |
| $K_j$ | | Equilibrium constant of reaction $j$ |
| $l_2$ | $= \dfrac{\text{Cathode channel length}}{\text{Anode channel length}}$ | Geometric aspect ratio of the cell |

*Molten Carbonate Fuel Cells.* Edited by Kai Sundmacher, Achim Kienle, Hans Josef Pesch,
Joachim F. Berndt, and Gerhard Huppmann
Copyright © 2007 WILEY-VCH Verlag GmbH & Co. KGaA, Weinheim
ISBN: 978-3-527-31474-4

**Table A.1** *(continued)*

**Latin symbols**

| | | |
|---|---|---|
| $n_j$ | | Numbers of electrons transferred in reaction $j$ |
| $Pe_s$ | $= \dfrac{\text{Characteristic convective heat transport}}{\text{Characteristic conductive heat transport}}$ | Peclet number |
| $Q_m$ | $= \dfrac{\text{Exchange heat flow}}{\text{Characteristic convective heat flow}}$ | Heat exchange flux |
| $q$ | $= \dfrac{\text{Exchange heat flow density}}{\text{Characteristic convective heat flow density}}$ | Heat exchange flux density |
| $q_s$ | $= \dfrac{\text{Heat source density}}{\text{Characteristic convective heat flow density}}$ | Heat source density |
| $r_j$ | $= \dfrac{\text{Reaction rate}}{\text{Characteristic reaction rate}}$ | Reaction rate of reaction $j$ |
| $R_{\text{back}}$ | | Cathode gas recycle ratio |
| $S/C$ | | Steam to carbon ratio |
| $St$ | $= \dfrac{\text{Characteristic heat exchange}}{\text{Characteristic convective heat transport}}$ | Stanton number |
| $U_{\text{cell}}$ | $= \dfrac{\text{Cell voltage}}{\text{Characteristic cell voltage}}$ | Cell voltage |
| $V$ | $= \dfrac{\text{Volume}}{\text{Characteristic volume}}$ | Volume |

**Greek symbols**

| | | |
|---|---|---|
| $\alpha_j$ | | Transition factors for electrochemical reaction $j$ |
| $\Gamma$ | $= \dfrac{\text{Molar flow}}{\text{Standard molar flow}}$ | Total molar flow |
| $\gamma$ | $= \dfrac{\text{Molar flow density}}{\text{Standard molar flow density}}$ | Molar flow density |
| $\Delta\phi_{j,0}$ | $= \dfrac{\text{Equilibrium potential difference}}{\text{Characteristic potential}}$ | Equilibrium potential difference of reaction $j$ |
| $\Delta_R c_{p,j}^0$ | $= \dfrac{\text{Molar heat capacity change}}{\text{Characteristic heat capacity}}$ | Heat capacity change due to reaction $j$ |
| $\Delta_R g_j^0$ | $= \dfrac{\text{Standard Gibbs enthalpy of reaction}}{\text{Characteristic sensible energy}}$ | Standard free (Gibbs) enthalpy of reaction $j$ |

**Table A.1** *(continued)*

**Greek symbols**

| | | |
|---|---|---|
| $\Delta_C h_i^0$ | $= \dfrac{\text{Standard enthalpy of combustion}}{\text{Characteristic sensible energy}}$ | Standard combustion enthalpy of component $i$ |
| $\Delta_f h_i^0$ | $= \dfrac{\text{Standard enthalpy of formation}}{\text{Characteristic sensible energy}}$ | Standard enthalpy of formation of component $i$ |
| $\Delta_R h_j^0$ | $= \dfrac{\text{Standard enthalpy of reaction}}{\text{Characteristic sensible energy}}$ | Standard enthalpy of reaction $j$ |
| $\Delta_R s_j^0$ | $= \dfrac{\text{Standard entropy of reaction}}{\text{Characteristic sensible energy}}$ | Standard entropy of reaction $j$ |
| $\zeta_1, \zeta_2$ | $= \dfrac{\text{Spatial coordinate}}{\text{Channel length}}$ | Spatial coordinate |
| $\vartheta$ | $= \dfrac{\text{Temperature}}{\text{Standard temperature}}$ | Temperature |
| $\kappa$ | $= \dfrac{\text{Characteristic ion flow density}}{\text{Characteristic current density}}$ | Ion conductivity |
| $\lambda_{air}$ | | Air number |
| $\nu_{i,j}$ | | Stoichiometric coefficient of component $i$ in reaction $j$ |
| $\nu_{i,j}^{\pm}$ | | Stoichiometric coefficient of educts/products in reaction $j$ |
| $\bar{\nu}_j$ | | Total mole change in reaction $j$ |
| $\nu_{i,Cj}$ | | Stoichiometric coefficient of component $i$ in combustion reaction of component $j$ |
| $\bar{\nu}_{Cj}$ | | Total mole change in combustion reaction of component $j$ |
| $\tau$ | $= \dfrac{\text{Time}}{\text{Characteristic residence time}}$ | Time |
| $\varphi_i$ | $= \dfrac{\text{Pressure}}{\text{Standard pressure}}$ | Partial pressure of component $i$ |
| $\phi$ | $= \dfrac{\text{Electric potential}}{\text{Characteristic electric potential}}$ | Electric potential |
| $\chi_i$ | | Mole fraction of component $i$ |

**Table A.2** Indexes

---

**Lower indexes**

---

| | |
|---|---|
| 0 | Initial condition |
| $\pm$ | Forward/backward reaction |
| $a$ | Anode gas phase |
| ac | Anode catalyst, anode electrode |
| air | Burner air |
| as | Anode/solid phase exchange |
| $b$ | Burner |
| back | Cathode gas recycle |
| $c$ | Cathode |
| cc | Cathode catalyst, cathode electrode |
| cell | Complete cell |
| cs | Cathode/solid phase exchange |
| $e$ | Electrolyte |
| $i$ | Chemical species, $i \in \{1 \dots 7\}$ |
| in | Inlet |
| iir | Indirect internal reformer |
| iirs | IIR/solid phase exchange |
| $j$ | Chemical reaction, $j \in \{\mathrm{iir1, iir2, ref1, ref2, ox1, ox2, red}\}$ |
| $m$ | Reversal chamber |
| out | Outlet |
| $s$ | Solid phase |

---

**Upper indexes**

---

| | |
|---|---|
| 0 | Thermodynamic standard |
| $L$ | Ion conducting phase of electrode |
| $S$ | Electron conducting phase of electrode |
| $\theta$ | System specific standard |

---

# B
# Benchmark Problem: Complete Set of Equations and Parameters

*Peter Heidebrecht*

In this section, the complete set of equations of the reference model is listed. The list of equations follows a certain order, but neither the equations nor the model assumptions will be discussed, as this has already been done in Chapter 3. A list of all model parameter values and a set of typical input conditions are also given. This should be suffice to implement this model in any type of software and solve it.

## B.1
## Equations

## B.2
## Parameters

The list of parameters is subdivided into constant model parameters and input conditions. The input conditions listed here correspond are imposed to the model in Chapter 6.

*Molten Carbonate Fuel Cells.* Edited by Kai Sundmacher, Achim Kienle, Hans Josef Pesch, Joachim F. Berndt, and Gerhard Huppmann
Copyright © 2007 WILEY-VCH Verlag GmbH & Co. KGaA, Weinheim
ISBN: 978-3-527-31474-4

**Table B.1** Model equations

---

*Indirect internal reformer*

Partial mass balance

$$\frac{V_{\text{iir}}}{\vartheta_{\text{iir}}} \cdot \frac{\partial \chi_{i,\text{iir}}}{\partial \tau} = \gamma_{\text{iir}} \cdot \frac{\partial \chi_{i,\text{iir}}}{\partial \zeta_1} + \sum_{j=\text{ref}} (v_{i,j} - \chi_{i,\text{iir}} \bar{v}_j) \cdot Da_{j,\text{iir}} r_j$$

Enthalpy balance

$$\frac{V_{\text{iir}}}{\vartheta_{\text{iir}}} \cdot \frac{\partial \vartheta_{\text{iir}}}{\partial \tau} = \gamma_{\text{iir}} \cdot \frac{\partial \vartheta_{\text{iir}}}{\partial \zeta_1} + \sum_{j=\text{ref}} \frac{-\Delta_R h_j}{\bar{c}_{p,\text{iir}}} \cdot Da_{j,\text{iir}} r_j + \frac{q_{\text{iirs}}}{\bar{c}_{p,\text{iir}}}$$

Total mass balance

$$0 = \frac{\partial \gamma_{\text{iir}} \vartheta_{\text{iir}}}{\partial \zeta_1} + \sum_{j=\text{ref}} \frac{-\Delta_R h_j}{\bar{c}_{p,\text{iir}}} \cdot Da_{j,\text{iir}} r_j + \frac{q_{\text{iirs}}}{\bar{c}_{p,\text{iir}}} + \vartheta_{\text{iir}} \cdot \sum_{j=\text{ref}} \bar{v}_j Da_{j,\text{iir}} r_j$$

Average heat capacity

$$\bar{c}_{p,\text{iir}} = \sum_i \chi_{i,\text{iir}} c_{p,i}$$

Average outlet values

$$\Gamma_{\text{iir,out}} = \int_0^1 \gamma_{\text{iir}}(\zeta_1 = 0, \zeta_2)\, d\zeta_2; \; \chi_{i,\text{iir,out}}$$

$$= \int_0^1 \frac{\gamma_{\text{iir}}(\zeta_1 = 0, \zeta_2)}{\Gamma_{\text{iir,out}}} \cdot \chi_{i,\text{iir}}(\zeta_1 = 0, \zeta_2)\, d\zeta_2; \; \vartheta_{\text{iir,out}}$$

$$= \vartheta^\theta + \int_0^1 \frac{\gamma_{\text{iir}}(\zeta_1 = 0, \zeta_2)}{\Gamma_{\text{iir,out}}} \cdot \frac{\sum_i \chi_{i,\text{iir}}(\zeta_1 = 0, \zeta_2) c_{p,i}}{\sum_i \chi_{i,\text{iir,out}} c_{p,i}}$$

$$\cdot (\vartheta_{\text{iir}}(\zeta_1 = 0, \zeta_2) - \vartheta^\theta)\, d\zeta_2$$

*Anode channel*

Partial mass balance

$$\frac{V_a}{\vartheta_a} \cdot \frac{\partial \chi_{i,a}}{\partial \tau} = -\gamma_a \cdot \frac{\partial \chi_{i,a}}{\partial \zeta_1} + \sum_{j=\text{ref,ox}} (v_{i,j} - \chi_{i,a} \bar{v}_j) \cdot Da_j r_j$$

Enthalpy balance

$$\frac{V_a}{\vartheta_a} \cdot \frac{\partial \vartheta_a}{\partial \tau} = -\gamma_a \cdot \frac{\partial \vartheta_a}{\partial \zeta_1} + \sum_{j=\text{ox}} \sum_i v_{i,j}^+ \cdot \frac{c_{p,i}}{\bar{c}_{p,a}}$$

$$\cdot Da_j r_j \cdot (\vartheta_s - \vartheta_a) + \sum_{j=\text{ref}} \frac{-\Delta_R h_j}{\bar{c}_{p,a}} \cdot Da_j r_j + \frac{q_{as}}{\bar{c}_{p,a}}$$

Total mass balance

$$0 = -\frac{\partial \gamma_a \vartheta_a}{\partial \zeta_1} + \sum_{j=\text{ox}} \sum_i v_{i,j}^+ \cdot \frac{c_{p,i}}{\bar{c}_{p,a}} \cdot Da_j r_j \cdot (\vartheta_s - \vartheta_a)$$

$$+ \sum_{j=\text{ref}} \frac{-\Delta_R h_j}{\bar{c}_{p,a}} \cdot Da_j r_j + \frac{q_{as}}{\bar{c}_{p,a}} + \vartheta_a \cdot \sum_{j=\text{ref,ox}} \bar{v}_j Da_j r_j$$

Average heat capacity

$$\bar{c}_{p,a} = \sum_i \chi_{i,a} c_{p,i}$$

Average outlet values

$$\Gamma_{a,\text{out}} = \int_0^1 \gamma_a(\zeta_1 = 1, \zeta_2)\, d\zeta_2; \; \chi_{i,a,\text{out}}$$

$$= \int_0^1 \frac{\gamma_a(\zeta_1 = 1, \zeta_2)}{\Gamma_{a,\text{out}}} \cdot \chi_{i,a}(\zeta_1 = 1, \zeta_2)\, d\zeta_2; \; \vartheta_{a,\text{out}}$$

$$= \vartheta^\theta + \int_0^1 \frac{\gamma_a(\zeta_1 = 1, \zeta_2)}{\Gamma_{a,\text{out}}} \cdot \frac{\sum_i \chi_{i,a}(\zeta_1 = 1, \zeta_2) c_{p,i}}{\sum_i \chi_{i,a,\text{out}} c_{p,i}}$$

$$\cdot (\vartheta_a(\zeta_1 = 1, \zeta_2) - \vartheta^\theta)\, d\zeta_2$$

**Table B.1** (continued)

*Burner*

Total mass balance
$$\Gamma_{b,\text{out}} = \Gamma_{a,\text{out}} \cdot \left(1 + \sum_j \bar{v}_{Cj} \cdot \chi_{j,a,\text{out}}\right)$$
$$+ \Gamma_{\text{back}} \cdot \left(1 + \sum_j \bar{v}_{Cj} \cdot \chi_{j,\text{back}}\right) + \Gamma_{\text{air}} \cdot \left(1 + \sum_j \bar{v}_{Cj} \cdot \chi_{j,\text{air}}\right)$$

Partial mass balance
$$\Gamma_{b,\text{out}} \cdot \chi_{i,b,\text{out}} = \Gamma_{a,\text{out}} \cdot \left(\chi_{i,a,\text{out}} + \sum_j v_{i,Cj} \cdot \chi_{j,a,\text{out}}\right)\Gamma_{\text{back}}$$
$$\cdot \left(\chi_{i,\text{back}} + \sum_j v_{i,Cj} \cdot \chi_{j,\text{back}}\right)\Gamma_{\text{air}}$$
$$\cdot \left(\chi_{i,\text{air}} + \sum_j v_{i,Cj} \cdot \chi_{j,\text{air}}\right)$$

Enthalpy balance
$$\Gamma_{b,\text{out}} \sum_i \chi_{i,b,\text{out}} \cdot c_{p,i} \cdot (\vartheta - \vartheta_{b,\text{out}})$$
$$= \Gamma_{a,\text{out}} \sum_i \chi_{i,a,\text{out}} \cdot (c_{p,i} \cdot (\vartheta_{a,\text{out}} - \vartheta^\theta) - \Delta_C h_i^0(\vartheta^\theta))$$
$$+ \Gamma_{\text{back}} \sum_i \chi_{i,\text{back}} \cdot (c_{p,i} \cdot (\vartheta_{\text{back}} - \vartheta^\theta) - \Delta_C h_i^0(\vartheta^\theta))$$
$$+ \Gamma_{\text{air}} \sum_i \chi_{i,\text{air}} \cdot (c_{p,i} \cdot (\vartheta_{\text{air}} - \vartheta^\theta) - \Delta_C h_i^0(\vartheta^\theta))$$

Combustion enthalpy
$$\Delta_C h_j^0(\vartheta) = \sum_i v_{i,Cj} \cdot \Delta_f h_i^0(\vartheta)$$

*Reversal chamber*

Partial mass balance
$$\frac{V_m}{\vartheta_m} \frac{d\chi_{i,m}}{d\tau} = \Gamma_{b,\text{out}} \cdot (\chi_{i,b,\text{out}} - \chi_{i,m})$$

Enthalpy balance
$$\frac{V_m \bar{c}_{p,m}}{\vartheta_m} \frac{d\vartheta_m}{d\tau} = \Gamma_{b,\text{out}} \cdot \bar{c}_{p,b,\text{out}} \cdot (\vartheta_{b,\text{out}} - \vartheta_m) + P_{\text{blower}} - Q_m$$

Total mass balance
$$\Gamma_m = \Gamma_{b,\text{out}} \cdot \left(1 + \frac{\bar{c}_{p,b,\text{out}}}{\bar{c}_{p,m}} \cdot \frac{\vartheta_{b,\text{out}} - \vartheta_m}{\vartheta_m}\right) - \frac{P_{\text{blower}} - Q_m}{\bar{c}_{p,m}\vartheta_m}$$

Heat loss
$$Q_m = St_m \cdot (\vartheta_m - \vartheta_u)$$

Average heat capacities
$$\bar{c}_{p,m} = \sum_i \chi_{i,m} \cdot c_{p,i}; \quad \bar{c}_{p,b,\text{out}} = \sum_i \chi_{i,b,\text{out}} \cdot c_{p,i}$$

*Cathode channel*

Partial mass balance
$$\frac{V_c}{\vartheta_c} \cdot \frac{\partial \chi_{i,c}}{\partial \tau} = -\gamma_c \cdot \frac{\partial \chi_{i,c}}{\partial \zeta_2} + \sum_{j=\text{red}} (v_{i,j} - \chi_{i,c}\bar{v}_j) \cdot Da_j r_j$$

Enthalpy balance
$$\frac{V_c}{\vartheta_c} \cdot \frac{\partial \vartheta_c}{\partial \tau} = -\gamma_c \cdot \frac{\partial \vartheta_c}{\partial \zeta_2} + \sum_{j=\text{red}} \sum_i v_{i,j}^- \cdot \frac{c_{p,i}}{\bar{c}_{p,c}} \cdot Da_j r_j \cdot (\vartheta_s - \vartheta_c) + \frac{q_{cs}}{\bar{c}_{p,c}}$$

Total mass balance
$$0 = -\frac{\partial \gamma_c \vartheta_c}{\partial \zeta_2} + \sum_{j=\text{red}} \sum_i v_{i,j}^- \cdot \frac{c_{p,i}}{\bar{c}_{p,c}} \cdot Da_j r_j \cdot (\vartheta_s - \vartheta_c)$$
$$+ \frac{q_{cs}}{\bar{c}_{p,c}} + \vartheta_c \cdot \sum_{j=\text{red}} \bar{v}_j Da_j r_j$$

**Table B.1** *(continued)*

---

Average heat capacity $\quad \bar{c}_{p,c} = \sum_i \chi_{i,c} c_{p,i}$

Average outlet values $\quad \Gamma_{c,\text{out}} = \int_0^1 \gamma_c(\zeta_1, \zeta_2 = 1)\, d\zeta_1;\ \chi_{i,c,\text{out}}$

$$= \int_0^1 \frac{\gamma_c(\zeta_1, \zeta_2 = 1)}{\Gamma_{c,\text{out}}} \cdot \chi_{i,c}(\zeta_1, \zeta_2 = 1)\, d\zeta_1;\ \vartheta_{c,\text{out}}$$

$$= \vartheta^\theta + \int_0^1 \frac{\gamma_c(\zeta_1, \zeta_2 = 1)}{\Gamma_{c,\text{out}}} \cdot \frac{\sum_i \chi_{i,c}(\zeta_1, \zeta_2 = 1) c_{p,i}}{\sum_i \chi_{i,c,\text{out}} c_{p,i}}$$

$$\cdot (\vartheta_c(\zeta_1, \zeta_2 = 1) - \vartheta^\theta)\, d\zeta_1$$

Cathode gas recycle $\quad \chi_{i,c,\text{out}} = \chi_{i,\text{exhaust}} = \chi_{i,\text{back}};\ \vartheta_{c,\text{out}} = \vartheta_{\text{exhaust}} = \vartheta_{\text{back}};$

$\quad \Gamma_{\text{back}} = R_{\text{back}} \cdot \Gamma_{c,\text{out}};\ \Gamma_{\text{exhaust}} = (1 - R_{\text{back}}) \cdot \Gamma_{c,\text{out}}$

*Electrode pores*

Partial mass balances $\quad 0 = \sum_{j=\text{ox}} v_{i,j} Da_j \cdot r_j - Di_{,\text{as}} \cdot (\varphi_{i,\text{as}} - \chi_{i,\text{as}});$

$\quad 0 = \sum_{j=\text{red}} v_{i,j} Da_j \cdot r_j - Di_{,\text{cs}} \cdot (\varphi_{i,\text{cs}} - \chi_{i,\text{cs}})$

*Solid phase*

Enthalpy balance $\quad c_{p,s} \dfrac{\partial \vartheta_s}{\partial \tau} = \dfrac{l_2}{Pe_s} \dfrac{\partial^2 \vartheta_s}{\partial \zeta_1^2} + \dfrac{1}{Pe_s l_2} \dfrac{\partial^2 \vartheta_s}{\partial \zeta_2^2} + \sum_{j=\text{ox}} \sum_i (-v_{i,j}^-) \cdot c_{p,i} \cdot Da_j r_j$

$$\cdot (\vartheta_a - \vartheta_s) + \sum_{j=\text{red}} \sum_i (-v_{i,j}^+) \cdot c_{p,i} \cdot Da_j r_j \cdot (\vartheta_c - \vartheta_s)$$

$$- q_{\text{as}} - q_{\text{cs}} - q_{\text{iirs}} + q_s$$

Heat source density $\quad q_s = \sum_{j=\text{ox}} (-\Delta_R h_j^0(\vartheta_s) + n_j(\phi_a^S - \phi_a^L)) Da_j r_j$

$$+ \sum_{j=\text{red}} (-\Delta_R h_j^0(\vartheta_s) + n_j(\phi_c^S - \phi_c^L)) Da_j r_j + (\phi_a^L - \phi_c^L) \cdot i_e \cdot \dfrac{1}{F}$$

Heat exchange densities $\quad q_{\text{as}} = St_{\text{as}} \cdot (\vartheta_s - \vartheta_a);\ q_{\text{cs}} = St_{\text{cs}} \cdot (\vartheta_s - \vartheta_c);\ q_{\text{iirs}} = St_{\text{iirs}} \cdot (\vartheta_s - \vartheta_i ir)$

*Electric potential*

Electric potentials $\quad \phi_a^S = 0;\ \dfrac{\partial \phi_a^L}{\partial \tau} = -\dfrac{1}{c_a} \cdot (i - i_a);\ \dfrac{\partial \phi_c^L}{\partial \tau} = -\dfrac{1}{c_a} \cdot (i - i_a) - \dfrac{1}{c_e} \cdot (i - i_e);$

$$\dfrac{\partial \phi_c^S}{\partial \tau} = \dfrac{I_a - I_{\text{cell}}}{c_a} + \dfrac{I_e - I_{\text{cell}}}{c_e} + \dfrac{I_c - I_{\text{cell}}}{c_c};\ U_{\text{cell}} = \phi_c^S - \phi_a^S$$

Current densities $\quad i_a(\phi_a^L) = \sum_{j=\text{ox}} n_j F Da_j r_j(\phi_a^L);$

$\quad i_c(\phi_c^L, \phi_c^S) = -\sum_{j=\text{red}} n_j F Da_j r_j(\phi_c^L, \phi_c^S);$

$\quad i_e(\phi_a^L, \phi_c^L) = \kappa_e \cdot (\phi_a^L - \phi_c^L);$

$\quad i = \left(\dfrac{1}{c_a} + \dfrac{1}{c_e} + \dfrac{1}{c_c}\right)^{-1} \cdot \left(\dfrac{i_a}{c_a} + \dfrac{i_e}{c_e} + \dfrac{i_c}{c_c} - \dfrac{I_a}{c_a} - \dfrac{I_e}{c_e} - \dfrac{I_c}{c_c}\right) + I_{\text{cell}}$

**Table B.1** (continued)

| | |
|---|---|
| Total currents | $I_a = \int_0^1 \int_0^1 i_a \, d\zeta_1 \, d\zeta_2; \; I_e = \int_0^1 \int_0^1 i_e \, d\zeta_1 \, d\zeta_2; \; I_c = \int_0^1 \int_0^1 i_c \, d\zeta_1 \, d\zeta_2$ |

*Reaction kinetics*

| | |
|---|---|
| Methane steam reforming | $r_{\mathrm{ref1}} = \exp\left( Arr_{\mathrm{ref1}} \cdot \left( \frac{1}{\vartheta^0} - \frac{1}{\vartheta} \right) \right) \cdot \left( \chi_{CH_4} \chi_{H_2O} - \frac{\chi_{CO} \chi_{H_2}^3}{K_{\mathrm{ref1}}(\vartheta)} \right)$ |
| Water gas shift reaction | $r_{\mathrm{ref2}} = \exp\left( Arr_{\mathrm{ref2}} \cdot \left( \frac{1}{\vartheta^0} - \frac{1}{\vartheta} \right) \right) \cdot \left( \chi_{CO} \chi_{H_2O} - \frac{\chi_{CO_2} \chi_{H_2}}{K_{\mathrm{ref2}}(\vartheta)} \right)$ |

Hydrogen oxidation

$$r_{\mathrm{ox1}} = \exp\left( Arr_{\mathrm{ox1}} \cdot \left( \frac{1}{\vartheta^0} - \frac{1}{\vartheta_s} \right) \right)$$

$$\cdot \left[ \varphi_{H_2,\mathrm{ac}} \cdot \exp\left( \alpha_{\mathrm{ox1},+} \cdot n_{\mathrm{ox1}} \frac{(\phi_a^S - \phi_a^L) - \Delta\phi_{\mathrm{ox1},0}(\vartheta_s)}{\vartheta_s} \right) \right.$$

$$- \varphi_{H_2O,\mathrm{ac}} \varphi_{CO_2,\mathrm{ac}}$$

$$\left. \cdot \exp\left( -(1 - \alpha_{\mathrm{ox1},+}) \cdot n_{\mathrm{ox1}} \frac{(\phi_a^S - \phi_a^L) - \Delta\phi_{\mathrm{ox1},0}(\vartheta_s)}{\vartheta_s} \right) \right]$$

Carbon monoxide oxidation

$$r_{\mathrm{ox2}} = \exp\left( Arr_{\mathrm{ox2}} \cdot \left( \frac{1}{\vartheta^0} - \frac{1}{\vartheta_s} \right) \right)$$

$$\cdot \left[ \varphi_{CO,\mathrm{ac}} \cdot \exp\left( \alpha_{\mathrm{ox2},+} \cdot n_{\mathrm{ox2}} \frac{(\phi_a^S - \phi_a^L) - \Delta\phi_{\mathrm{ox2},0}(\vartheta_s)}{\vartheta_s} \right) \right.$$

$$\left. - \varphi_{CO_2,\mathrm{ac}}^2 \cdot \exp\left( -(1 - \alpha_{\mathrm{ox2},+}) \cdot n_{\mathrm{ox2}} \frac{(\phi_a^S - \phi_a^L) - \Delta\phi_{\mathrm{ox2},0}(\vartheta_s)}{\vartheta_s} \right) \right]$$

Oxygen reduction

$$r_{\mathrm{red}} = \exp\left( Arr_{\mathrm{red}} \cdot \left( \frac{1}{\vartheta^0} - \frac{1}{\vartheta_s} \right) \right)$$

$$\cdot \left[ \varphi_{CO_2,\mathrm{cc}}^{-2} \cdot \exp\left( 2.5 \cdot \frac{(\phi_c^S - \phi_c^L) - \Delta\phi_{\mathrm{red},0}(\vartheta_s)}{\vartheta_s} \right) \right.$$

$$\left. - \varphi_{O_2,\mathrm{cc}}^{3/4} \cdot \varphi_{CO_2,\mathrm{cc}}^{-1/2} \cdot \exp\left( -0.5 \frac{(\phi_c^S - \phi_c^L) - \Delta\phi_{\mathrm{red},0}(\vartheta_s)}{\vartheta_s} \right) \right]$$

*Thermodynamics*

| | |
|---|---|
| Free enthalpy of reaction | $\Delta_R g_j^0(\vartheta) = \Delta_R h_j^0(\vartheta^0) - \vartheta \cdot \Delta_R s_j^0(\vartheta^0)$ |
| Reaction enthalpy/ entropy | $\Delta_R h_j^0(\vartheta) = \sum_i \nu_{i,j} \cdot \Delta_f h_i^0(\vartheta); \; \Delta_R s_j^0(\vartheta) = \sum_i \nu_{i,j} \cdot s_{f,i}^0(\vartheta)$ |
| Temperature dependent enthalpy | $\Delta_R h_j^0(\vartheta) = \Delta_R h_j^0(\vartheta^0) + \Delta_R c_{p,j}^0(\vartheta^0) \cdot (\vartheta - \vartheta^0);$ $\Delta_R c_{p,j}^0 = \sum_i \nu_{i,j} \cdot c_{p,i}$ |

**Table B.2** Parameter values

---

System parameters: Reaction kinetics

---

|            | iir1  | iir2  | ref1  | ref2  | ox1   | ox2   | red   |
| ---------- | ----- | ----- | ----- | ----- | ----- | ----- | ----- |
| $Da_j$     | 200   | 200   | 200   | 200   | 6.8   | 0.0   | 0.121 |
| $Arr_j$    | 84.4  | 6.2   | 84.4  | 6.2   | 21.6  | 21.6  | 31.2  |
| $\alpha_{j,+}$ | –   | –     | –     | –     | 0.5   | 0.5   | –     |
| $\alpha_{j,-}$ | –   | –     | –     | –     | 0.5   | 0.5   | –     |

---

System parameters: Thermodynamics

---

|                           | $CH_4$ | $H_2O$ | $H_2$ | $CO$   | $CO_2$   | $O_2$ | $N_2$ |
| ------------------------- | ------ | ------ | ----- | ------ | -------- | ----- | ----- |
| $\Delta_f h_i^0(\vartheta^0)$ | −18.36 | −88.96 | 6.82 | −37.35 | −147.69  | 7.39  | 7.12  |
| $s_{f,i}^0(\vartheta^0)$  | 28.65  | 27.44  | 19.50 | 27.78 | 31.75    | 28.74 | 27.01 |
| $c_{p,i}$                 | 7.92   | 4.89   | 3.63  | 3.94  | 6.38     | 4.06  | 3.89  |
| $\vartheta^0$             | 2.929  |        |       |       |          |       |       |

---

System parameters: Stoichiometry

---

| $\nu_{i,j}$ |         | iir1        | iir2       | ref1       | ref2       | ox1        | ox2        | red        |
| ----------- | ------- | ----------- | ---------- | ---------- | ---------- | ---------- | ---------- | ---------- |
|             | $CH_4$  | −1          | 0          | −1         | 0          | 0          | 0          | 0          |
|             | $H_2O$  | −1          | −1         | −1         | −1         | 1          | 0          | 0          |
|             | $H_2$   | 3           | 1          | 3          | 1          | −1         | 0          | 0          |
|             | $CO$    | 1           | −1         | 1          | −1         | 0          | −1         | 0          |
|             | $CO_2$  | 0           | 1          | 0          | 1          | 1          | 2          | −1         |
|             | $O_2$   | 0           | 0          | 0          | 0          | 0          | 0          | −0.5       |
|             | $N_2$   | 0           | 0          | 0          | 0          | 0          | 0          | 0          |
| $n_j$       |         | –           | –          | –          | –          | 2          | 2          | 2          |
| $\nu_{i,Cj}$ |        | $C_{CH_4}$  | $C_{H_2O}$ | $C_{H_2}$  | $C_{CO}$   | $C_{CO_2}$ | $C_{O_2}$  | $C_{N_2}$  |
|             | $CH_4$  | −1          | 0          | 0          | 0          | 0          | 0          | 0          |
|             | $H_2O$  | 2           | 0          | 1          | 0          | 0          | 0          | 0          |
|             | $H_2$   | 0           | 0          | −1         | 0          | 0          | 0          | 0          |
|             | $CO$    | 0           | 0          | 0          | −1         | 0          | 0          | 0          |
|             | $CO_2$  | 1           | 0          | 0          | 1          | 0          | 0          | 0          |
|             | $O_2$   | −2          | 0          | −0.5       | −0.5       | 0          | 0          | 0          |
|             | $N_2$   | 0           | 0          | 0          | 0          | 0          | 0          | 0          |
| $F$         | 3.5/8   |             |            |            |            |            |            |            |

---

System parameters: Heat transfer

---

|              | iirs | as  | cs  | m   |
| ------------ | ---- | --- | --- | --- |
| $St$         | 38   | 40  | 138 | 0.5 |
| $Pe_s$       | 2.5  |     |     |     |
| $\vartheta_u$ | 1.0 |     |     |     |

**Table B.2** *(continued)*

System parameters: Accumulation

| $c_{p,s}$ | 10,000 | | |
|---|---|---|---|
| | iir | a | c |
| $V$ | 1.0 | 1.0 | 1.0 |
| | a | e | c |
| $c$ | $1.0 \times 10^{-5}$ | $1.0 \times 10^{-5}$ | $1.0 \times 10^{-5}$ |

System parameters: Accumulation

| $D_{i,as}$ | 100 |
|---|---|
| $D_{i,cs}$ | 100 |
| $\kappa_e$ | 1 |

System parameters: Others

| $l_2$ | 0.666 |
|---|---|

**Table B.3** Input conditions

Input parameters: Fuel gas

|  | $CH_4$ | $H_2O$ | $H_2$ | $CO$ | $CO_2$ | $O_2$ | $N_2$ |
|---|---|---|---|---|---|---|---|
| $\chi_{i, \text{iir, in}}$ | 0.222 | 0.574 | 0.151 | 0.0 | 0.047 | 0.0 | 0.006 |
| $\Gamma_{\text{iir, in}}$ | 0.826 |  |  |  |  |  |  |
| $\vartheta_{\text{iir, in}}$ | 2.170 |  |  |  |  |  |  |

Input parameters: Air

|  | $CH_4$ | $H_2O$ | $H_2$ | $CO$ | $CO_2$ | $O_2$ | $N_2$ |
|---|---|---|---|---|---|---|---|
| $\chi_{i, \text{air}}$ | 0 | 0 | 0 | 0 | 0 | 0.21 | 0.79 |
| $\lambda_{\text{air}}$ | 2.69 |  |  |  |  |  |  |
| $\vartheta_{\text{air}}$ | 1.076 |  |  |  |  |  |  |

Input parameters: Others

| $I_{\text{cell}}$ | 0.4473 |
|---|---|
| $R_{\text{back}}$ | 0.7 |
| $P_{\text{blower}}$ | 2.775 |

# Index

*Molten Carbonate Fuel Cells.* Kai Sundmacher, Achim Kienle, Hans Josef Pesch, Joachim F. Berndt,
and Gerhard Huppmann (Eds.)
Copyright © 2007 WILEY-VCH Verlag GmbH & Co. KGaA, Weinheim
ISBN: 978-3-527-31474-4

## Further Reading

Sundmacher, K., Kienle, A., Seidel-Morgenstern, A. (Eds.)

**Integrated Chemical Processes**

**Syntheses, Operation, Analysis, and Control**

2005
Hardcover
ISBN: 978-3-527-30831-6

Olah, G. A., Goeppert, A., Prakash, G. K. S.

**Beyond Oil and Gas: The Methanol Economy**

2006
Hardcover
ISBN: 978-3-527-31275-7

Elvers, B. (Ed.)

**Handbook of Fuels**

**Energy Sources for Transportation**

2007
Hardcover
ISBN: 978-3-527-30740-1

Häring, W. (Ed.)

**Industrial Gases Processing**

2007
Hardcover
ISBN: 978-3-527-31685-4

Vielstich, W., Lamm, A., Gasteiger, H. (Eds.)

**Handbook of Fuel Cells**

**Fundamentals, Technology, Applications**

4 Volumes
2003
Hardcover
ISBN: 978-0-471-49926-8